Applied Probability Theory - New Perspectives, Recent Advances and Trends

Edited by Abdo Abou Jaoudé

Published in London, United Kingdom

Applied Probability Theory - New Perspectives, Recent Advances and Trends
http://dx.doi.org/10.5772/intechopen.104026
Edited by Abdo Abou Jaoudé

Contributors
Tatiana K. Kronberg, Alexander A. Kronberg, Saul Antonio Obregón Biosca, José Luis Reyes Araiza, Miguel Angel Pérez Lara y Hernández, José Alfredo Jiménez Moscoso, Gerson Yahir Palomino Velandia, Sandile Shongwe, Retsebile Maphalla, Moroke Mokhoabane, Mulalo Ndou, Abdo Abou Jaoudé

First published in London, United Kingdom, 2023 by IntechOpen
IntechOpen is the global imprint of INTECHOPEN LIMITED, registered in England and Wales, registration number: 11086078, 5 Princes Gate Court, London, SW7 2QJ, United Kingdom

British Library Cataloguing-in-Publication Data
A catalogue record for this book is available from the British Library

Additional hard and PDF copies can be obtained from orders@intechopen.com

Applied Probability Theory - New Perspectives, Recent Advances and Trends
Edited by Abdo Abou Jaoudé
p. cm.
Print ISBN 978-1-83768-295-9
Online ISBN 978-1-83768-296-6
eBook (PDF) ISBN 978-1-83768-297-3

Meet the editor

Abdo Abou Jaoudé has been teaching for many years and has a passion for researching mathematics. He is currently an Associate Professor of Mathematics and Statistics at Notre Dame University-Louaizé (NDU), Lebanon. He holds a BSc and an MSc in Computer Science from NDU, and three Ph.D.`s in Applied Mathematics, Computer Science, and Applied Statistics and Probability from Bircham International University, Spain. He also holds two PhDs in Mathematics and Prognostics from the Lebanese University, Lebanon, and Aix-Marseille University, France. Dr. Abou Jaoudé's broad research interests include pure and applied mathematics. He has published twenty-three international journal articles and six contributions to conference proceedings, in addition to nine books on prognostics, mathematics, and computer science.

Contents

Preface

Probability theory is a branch of statistics that employs mathematical methods of collection, organization, and interpretation of data, with applications in practically all scientific areas. When working with probability theory, we analyze random phenomena and assess the likelihood that an event will occur. This book, *Applied Probability Theory - New Perspectives, Recent Advances and Trends*, discusses some fundamental aspects of probability theory and explores its use to solve a large array of problems. Chapters address such topics as complex probability, the stability of algorithms in statistical modeling, the non-homogeneous Hofmann process, and more.

Each time I work in the field of mathematical probability and statistics I find pleasure in tackling the knowledge, theorems, proofs, and applications of the theory. Each problem is a riddle to be solved and I become relieved and extremely happy when I reach the riddle's solution. This proves two important facts: first, the power of mathematics and its models to deal with such kinds of problems and second, the power of the human mind to understand such problems and tame the wild concepts of randomness, probability, stochasticity, uncertainty, chaos, chance, and non-determinism.

I chose the word paradigm for this branch of mathematical sciences after consulting the influential book *The Structure of Scientific Revolutions* by Thomas Kuhn, in which the author used the term to describe a set of theories, standards, and methods that together represent a way of organizing knowledge, that is, a model or a way of viewing the world. Kuhn stated in his thesis that revolutions in science occur when an older paradigm is reexamined, rejected, and replaced by another, just like Einstein's theories of special and general relativity that dethroned Newtonian mechanistic theory, or quantum mechanics that replaced the classical theories of electromagnetism and thermodynamics when probing the micro-world. What about probability and statistics? We can affirm that their set of theories and methods developed across the centuries have defined for us a way to view the world and a model to understand and deal with such concepts as randomness, chance, stochasticity, chaos, probability, and so on. Hence, the definition of a paradigm suits very well this discipline of knowledge and this methodology of thinking. This justifies my usage of this term in my two chapters of this book.

I hope that after reading this book you will recognize my amazement and wonder at the power of the theory of probability and statistics to deal with randomness, as well as my excitement to delve into the depths of a very profound field in mathematics. Thus, to convey my impression of wonder I cite the following words of Albert Einstein:

"The most incomprehensible thing about the universe is that it is comprehensible..."

Furthermore, although I have taught courses on probability and statistics at the university level for many years, I consider myself a beginner in this branch of knowledge; in fact an *absolute beginner*, always thirsty to learn and discover more. I think that the mathematician who proves to be successful in tackling and mastering the theory of probability and statistics has made it halfway to understanding the mystery of existence revealed in a universe governed sometimes in our modern theories by randomness and uncertainties. The probabilistic aspect is evident in the theories of the quantum world, of thermodynamics, or of statistical mechanics, for example. Hence, the universe's secret code, I think, is written in a mathematical language, just as Galileo Galilei expressed it in these words:

"Philosophy is written in this very great book which is the universe that always lies open before our eyes. One cannot understand this book unless one first learns to understand the language and recognize the characters in which it is written. It is written in a mathematical language and the characters are triangles, circles and other geometrical figures. Without these means it is humanly impossible to understand a word of it. Without these there is only clueless scrabbling around in a dark labyrinth."

Some may criticize my opinion and say that the theory of probability and statistics is a speculative and an uncertain science dealing with approximations and uncertainties. That is completely true. But since this field, or paradigm, is a part of mathematics, it has allowed us to understand, measure quantitatively, and tame chaos, even if not completely and absolutely. In fact, probability theory keeps the spirit and the flavor of "exact" sciences through its numbers, proofs, figures, theorems, and graphs.

To conclude, I am truly astonished by the power of probability theory to deal with random data and phenomena, and this feeling and impression have never left me since the first time I was introduced to this branch of science and mathematics. I hope that this book will convey and share this feeling with readers.

Abdo Abou Jaoudé, Ph.D.
Notre Dame University-Louaizé,
Zouk Mosbeh, Lebanon

Chapter 1

The Paradigm of Complex Probability and Quantum Mechanics: The Infinite Potential Well Problem – The Position Wave Function

Abdo Abou Jaoudé

Abstract

The system of axioms for probability theory laid in 1933 by Andrey Nikolaevich Kolmogorov can be extended to encompass the imaginary set of numbers and this by adding to his original five axioms an additional three axioms. Therefore, we create the complex probability set $\mathcal{C}$, which is the sum of the real set $\mathcal{R}$ with its corresponding real probability, and the imaginary set $\mathcal{M}$ with its corresponding imaginary probability. Hence, all stochastic experiments are performed now in the complex set $\mathcal{C}$ instead of the real set $\mathcal{R}$. The objective is then to evaluate the complex probabilities by considering supplementary new imaginary dimensions to the event occurring in the "real" laboratory. Consequently, the corresponding probability in the whole set $\mathcal{C}$ is always equal to one and the outcome of the random experiments that follow any probability distribution in $\mathcal{R}$ is now predicted totally in $\mathcal{C}$. Subsequently, it follows that chance and luck in $\mathcal{R}$ is replaced by total determinism in $\mathcal{C}$. Consequently, by subtracting the chaotic factor from the degree of our knowledge of the stochastic system, we evaluate the probability of any random phenomenon in $\mathcal{C}$. My innovative complex probability paradigm (*CPP*) will be applied to the established theory of quantum mechanics in order to express it completely deterministically in the universe $\mathcal{C} = \mathcal{R} + \mathcal{M}$.

Keywords: chaotic factor, degree of our knowledge, complex random vector, probability norm, complex probability set $\mathcal{C}$, position wave function

> *"Nothing in nature is by chance … Something appears to be chance only because of our lack of knowledge."*
>
> *Baruch Spinoza.*
>
> *"You believe in the God who plays dice, and I in complete law and order."*
>
> *Albert Einstein, Letter to Max Born*
>
> *"Without mathematics, we cannot penetrate deeply into philosophy.*
> *Without philosophy, we cannot penetrate deeply into mathematics.*
> *Without both, we cannot penetrate deeply into anything… ".*
>
> *Gottfried Wilhelm von Leibniz.*

"There are more things in heaven and earth, Horatio, than are dreamt of in your philosophy."

Hamlet (1601), William Shakespeare

1. Introduction

There are several names for this idealized and highly artificial potential, prominent among them: *The infinite square-well potential*, *the infinite potential* [1–3]. However, it is the phrase *particle in an escape-proof box* that is more likely to be intuitively appealing. Very simply it tells us about a particle moving inside the box as a *free particle* except at the box walls, which are postulated to be impenetrable by definition. The particle in an escape proof box is a sleek, easy-on-the-mathematics model for initiating students into quantum mechanics, with the added advantage that it is one of the few sectors within quantum mechanics where the Schrödinger equation can be solved analytically without resorting to approximation techniques. In the context of this simple potential, students typically find their first very intuitive understanding of the meaning of bound states, boundary conditions, stationary states, and energy-momentum quantization. It is even an introduction to quantum tunneling by emphasizing by contrast *why a particle in a box cannot tunnel out of the box!* The infinite square well is then an easy introduction to a more general understanding of the time independent Schrödinger equation for bound states in more sophisticated potentials, *where the quantum tunneling phenomenon is exhibited.* It is easy stepping-stones away from this *first* potential to the more complicated structures, such as the simple harmonic oscillator, which plays a seminal role in quantum field theory. In a clear and present sense, the quantum adventure can fairly be said to begin with this humble but very remarkable particle in an escape-proof box conception. However, *genius in simplicity* is another watchword for this potential. Remarkably, from such simplicity, one is also able to extract an enormous amount of excellent physics. Never mind that there are no actual confining forces in the world that are infinitely strong, physicists successfully deploy the square well potential to model complicated physics all the time, witness the infinite square well potential, which was used by physicist Sommerfeld to model his electron gas theory, where he construed the moving electrons as free particles confined to an escape-proof box. And again, the particle in a box is also deployed to model and investigate a myriad of other complex physical systems – the Hexatriene molecule, among others, as well as in fabricated semiconductor layers. As Cartwright notes, "Of course, this is not a true description of the potentials that are actually produced by the walls and the environment. But is not exactly false either. It is just the way to achieve the results in the model that the walls and environment are supposed to achieve in reality. The infinite potential is a good piece of stage setting." [3] True, the particle in the escape proof-box is by definition a highly contrived and idealized model. Consequently, this important and well-known problem in quantum mechanics will be related to my complex probability paradigm (*CPP*) in order to express it totally deterministically.

In the end, and to conclude, this research work's first chapter is organized as follows: After the introduction in section 1, the purpose and the advantages of the present work are presented in section 2. Afterward, in section 3, the extended Kolmogorov's axioms, and hence, the complex probability paradigm with their original parameters and interpretation, will be explained and summarized. Moreover, in section 4, we will explain briefly the one-dimensional case of the infinite square well problem considered in this work. Additionally, in section 5, the new paradigm will be related to the particle in a box problem after applying *CPP* to the position wave function; hence, some corresponding simulations will be done, and afterward, the characteristics of this stochastic distribution will be computed in the probabilities sets $\mathcal{R}$, $\mathcal{M}$, and $\mathcal{C}$. Finally, we conclude the work by doing a comprehensive summary in section 6, and then present in section 7 the list of references cited in the current chapter. Furthermore, in the following second chapter, the new paradigm will be related to the particle in a box problem after applying *CPP* to the momentum wave function of the problem; hence, some corresponding simulations will be done, and afterward, the characteristics of this stochastic distribution will be computed in the probabilities sets $\mathcal{R}$, $\mathcal{M}$, and $\mathcal{C}$. Also, in the following chapter, *CPP* will be used to extend and verify Heisenberg uncertainty principle in $\mathcal{R}$, $\mathcal{M}$, and $\mathcal{C}$. In addition, we will calculate and determine the position and the momentum wave functions entropies in $\mathcal{R}$, $\mathcal{M}$, and $\mathcal{C}$.

2. The purpose and the advantages of the current publication

All our work in classical probability theory is to compute probabilities. The original idea in this research work is to add new dimensions to our random experiment, which will make the work deterministic. In fact, the probability theory is a nondeterministic theory by nature, that means that the outcome of the events is due to chance and luck. By adding new dimensions to the event in $\mathcal{R}$, we make the work deterministic, and hence, a random experiment will have a certain outcome in the complex set of probabilities $\mathcal{C}$. It is of great importance that the stochastic system, like the problem in quantum mechanics considered here, becomes totally predictable since we will be totally knowledgeable to foretell the outcome of chaotic and random events that occur in nature, for example, in statistical mechanics or in all stochastic processes. Therefore, the work that should be done is to add to the real set of probabilities $\mathcal{R}$, the contributions of $\mathcal{M}$, which is the imaginary set of probabilities that will make the event in $\mathcal{C} = \mathcal{R} + \mathcal{M}$ deterministic. If this is found to be fruitful, then a new theory in statistical sciences and prognostic, and mainly in quantum mechanics is elaborated, this is to understand absolutely deterministically those phenomena that used to be random phenomena in $\mathcal{R}$. This is what I called "The Complex Probability Paradigm (*CPP*)," which was initiated and elaborated in my 19 previous papers [4–22].

To summarize, the advantages and the purposes of this current work are to:

1. Extend the theory of classical probability to encompass the complex numbers set, hence to bond the theory of probability to the field of complex variables and analysis in mathematics. This mission was elaborated on and initiated in my earlier 19 papers.

2. Apply the novel probability axioms and *CPP* paradigm to quantum mechanics, specifically to the infinite potential well problem.

3. Show that all nondeterministic phenomena like in the problem considered here can be expressed deterministically in the complex probabilities set $\mathcal{C}$.

4. Compute and quantify both the degree of our knowledge and the chaotic factor of the wave function position and momentum distributions and *CPP* in the sets $\mathcal{R}$, $\mathcal{M}$, and $\mathcal{C}$.

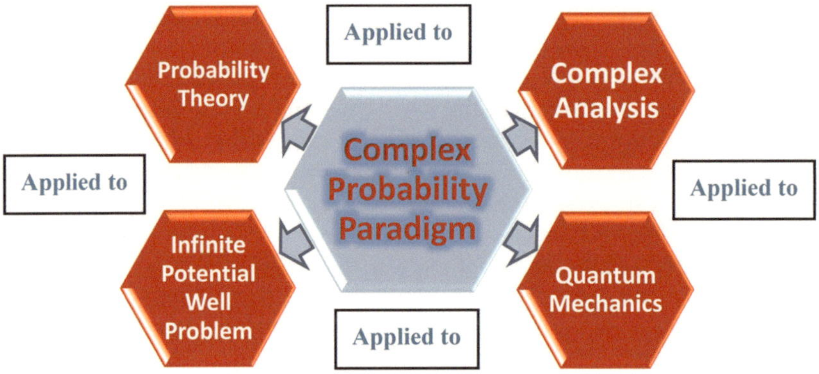

Figure 1.
The diagram of the complex probability paradigm applied to quantum mechanics major purposes and goals.

5. Represent and show the graphs of the functions and parameters of the innovative paradigm related to quantum mechanics.

6. Evaluate all the characteristics of the wave function position and momentum distributions.

7. Demonstrate that the classical concept of probability is permanently equal to one in the set of complex probabilities, hence, no randomness, no chaos, no ignorance, no uncertainty, no nondeterminism, and no unpredictability exist in:

$$\mathcal{C} \text{ (complex set)} = \mathcal{R} \text{ (real set)} + \mathcal{M} \text{ (imaginary set)}.$$

8. Calculate the problem entropies in $\mathcal{R}$, $\mathcal{M}$, and $\mathcal{C}$, and show that there is no disorder and no information loss nor gain in *CPP* but conservation of information.

9. Verify and extend Heisenberg uncertainty principle in $\mathcal{R}$ to $\mathcal{M}$ and $\mathcal{C}$.

10. Prepare to implement this creative model to other topics and problems in quantum mechanics. These will be the job to be accomplished in my future research publications.

Concerning some applications of the novel founded paradigm and as future work, it can be applied to any nondeterministic phenomenon in quantum mechanics. And compared with existing literature, the major contribution of the current research work is to apply the innovative paradigm of *CPP* to quantum mechanics and to express it completely deterministically. The next figure displays the major purposes of the complex probability paradigm (*CPP*) (**Figure 1**).

3. The complex probability paradigm

3.1 The original Andrey Nikolaevich Kolmogorov system of axioms

The simplicity of Kolmogorov's system of axioms may be surprising [4–22]. Let E be a collection of elements $\{E_1, E_2, \ldots\}$ called elementary events, and let F be a set of subsets of E called random events [23–27]. The five axioms for a finite set E are:

Axiom 1: F is a field of sets.
Axiom 2: F contains the set E.
Axiom 3: A nonnegative real number $P_{rob}(A)$ called the probability of A, is assigned to each set A in F. We have always $0 \le P_{rob}(A) \le 1$.
Axiom 4: $P_{rob}(E)$ equals 1.
Axiom 5: If A and B have no elements in common, the number assigned to their union is:

$$P_{rob}(A \cup B) = P_{rob}(A) + P_{rob}(B)$$

hence, we say that A and B are disjoint; otherwise, we have:

$$P_{rob}(A \cup B) = P_{rob}(A) + P_{rob}(B) - P_{rob}(A \cap B)$$

And we say also that: $P_{rob}(A \cap B) = P_{rob}(A) \times P_{rob}(B/A) = P_{rob}(B) \times P_{rob}(A/B)$ which is the conditional probability. If both A and B are independent then: $P_{rob}(A \cap B) = P_{rob}(A) \times P_{rob}(B)$.

Moreover, we can generalize and say that for N disjoint (mutually exclusive) events $A_1, A_2, \dots, A_j, \dots, A_N$ (for $1 \le j \le N$), we have the following additivity rule:

$$P_{rob}\left(\bigcup_{j=1}^{N} A_j\right) = \sum_{j=1}^{N} P_{rob}(A_j)$$

And we say also that for N independent events $A_1, A_2, \dots, A_j, \dots, A_N$ (for $1 \le j \le N$), we have the following product rule:

$$P_{rob}\left(\bigcap_{j=1}^{N} A_j\right) = \prod_{j=1}^{N} P_{rob}(A_j)$$

3.2 Adding the imaginary part $\mathcal{M}$

Now, we can add to this system of axioms an imaginary part such that:
Axiom 6: Let $P_m = i \times (1 - P_r)$ be the probability of an associated complementary event in $\mathcal{M}$ (the imaginary part or universe) to the event A in $\mathcal{R}$ (the real part or universe). It follows that $P_r + P_m/i = 1$, where i is the imaginary number with $i = \sqrt{-1}$ or $i^2 = -1$.
Axiom 7: We construct the complex number or vector $Z = P_r + P_m = P_r + i(1 - P_r)$ having a norm $|Z|$ such that:

$$|Z|^2 = P_r^2 + (P_m/i)^2.$$

Axiom 8: Let Pc denote the probability of an event in the complex probability set and universe $\mathcal{C}$, where $\mathcal{C} = \mathcal{R} + \mathcal{M}$. We say that Pc is the probability of an event A in $\mathcal{R}$ with its associated and complementary event in $\mathcal{M}$ such that:

$$Pc^2 = (P_r + P_m/i)^2 = |Z|^2 - 2iP_rP_m \text{ and is always equal to 1.}$$

We can see that by taking into consideration the set of imaginary probabilities we added three new and original axioms and consequently the system of axioms

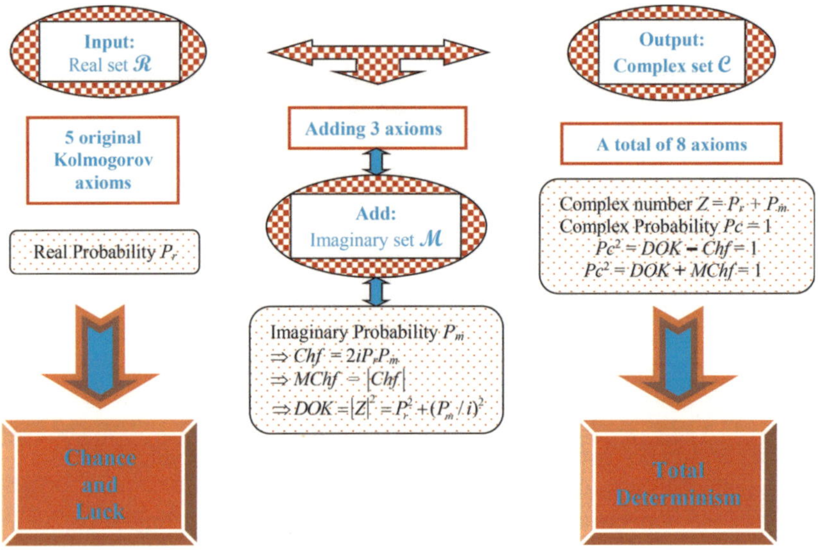

Figure 2.
The EKA *or the* CPP *diagram.*

defined by Kolmogorov was hence expanded to encompass the set of imaginary numbers and realm [28–65].

3.3 A concise interpretation of the original *CPP* paradigm

To summarize the novel *CPP* paradigm, we state that in the real probability universe $\mathcal{R}$ the degree of our certain knowledge is undesirably imperfect, and hence, unsatisfactory, thus we extend our analysis to the set of complex numbers $\mathcal{C}$, which incorporates the contributions of both the set of real probabilities, which is $\mathcal{R}$ and the complementary set of imaginary probabilities, which is $\mathcal{M}$. Afterward, this will yield an absolute and perfect degree of our knowledge in the probability universe $\mathcal{C} = \mathcal{R} + \mathcal{M}$ because $Pc = 1$ constantly and permanently. As a matter of fact, the work in the universe $\mathcal{C}$ of complex probabilities gives way to a sure forecast of any stochastic experiment, since in $\mathcal{C}$ we remove and subtract from the computed degree of our knowledge the measured chaotic factor. This will generate in universe $\mathcal{C}$ a probability equal to 1 $(Pc^2 = DOK - Chf = DOK + MChf = 1 = Pc)$. Many applications which take into consideration numerous continuous and discrete probability distributions in my 19 previous research papers confirm this hypothesis and innovative paradigm [4–22]. The Extended Kolmogorov Axioms (*EKA* for short) or the Complex Probability Paradigm (*CPP* for short) can be shown and summarized in the next illustration (**Figure 2**):

4. One-dimensional case of the infinite potential well problem

The simplest form of the particle in a box model considers a one-dimensional system [1, 2]. Here, the particle may only move backward and forwards along a straight line with impenetrable barriers at either end. The walls of a one-dimensional box may be seen as regions of space with an infinitely large potential energy. Conversely, the interior of the box has a constant zero potential energy. This means that

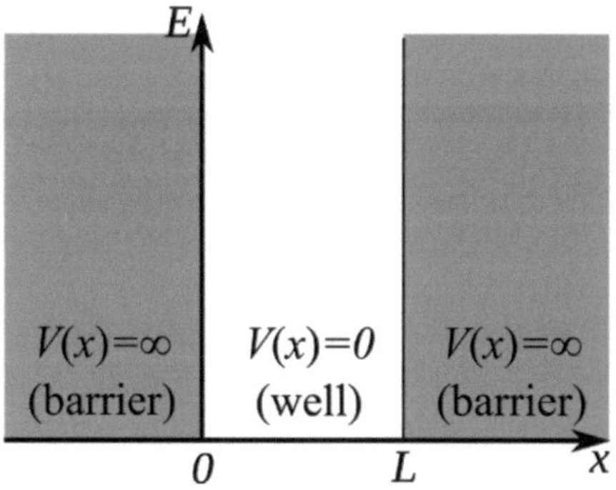

Figure 3.
The barriers outside a one-dimensional box have infinitely large potential, while the interior of the box has a constant zero potential.

no forces act upon the particle inside the box and it can move freely in that region. However, infinitely large forces repel the particle if it touches the walls of the box, preventing it from escaping. The potential energy in this model is given as:

$$V(x) = \begin{cases} 0 & x_c - \dfrac{L}{2} < x < x_c + \dfrac{L}{2} \\ \infty & \text{otherwise} \end{cases}$$

where L is the length of the box, x_c is the location of the center of the box and x is the position of the particle within the box. Simple cases include the centered box ($x_c = 0$) and the shifted box ($x_c = \frac{L}{2}$) (**Figure 3**).

5. The infinite potential well problem in quantum mechanics and the complex probability paradigm (*CPP*) parameters

In this section, we will relate and link quantum mechanics to the complex probability paradigm with all its parameters by applying it to the infinite potential well problem and by using the four *CPP* concepts which are: the real probability P_r in the real probability set $\mathcal{R}$, the imaginary probability P_m in the imaginary probability set $\mathcal{M}$, the complex random vector or number Z in the complex probability set $\mathcal{C} = \mathcal{R} + \mathcal{M}$, and the deterministic real probability Pc also in the probability set $\mathcal{C}$ [1–22, 66–99].

5.1 The position wave function and *CPP*: The position wave function solution

In quantum mechanics, the wave function gives the most fundamental description of the behavior of a particle; the measurable properties of the particle (such as its position, momentum, and energy) may all be derived from the wave function. The wave function $\psi(x, t)$ can be found by solving the Schrödinger equation for the system:

$$i\hbar\frac{\partial}{\partial t}\psi(x, t) = -\frac{\hbar^2}{2m}\frac{\partial^2}{\partial x^2}\psi(x, t) + V(x)\psi(x, t)$$

where $\hbar = \frac{h}{2\pi}$ is the reduced Planck constant, m is the mass of the particle, i is the imaginary unit, and t is time.

Inside the box, no forces act upon the particle, which means that the part of the wave function inside the box oscillates through space and time in the same form as a free particle:

$$\psi(x,t) = [A\sin(kx) + B\cos(kx)]e^{-i\omega t}$$

where A and B are arbitrary complex numbers. The frequency of the oscillations through space and time is given by the wave number k and the angular frequency ω, respectively.

$$\Leftrightarrow \psi_n(x,t) = \begin{cases} A\sin\left[k_n\left(x - x_c + \frac{L}{2}\right)\right]e^{-i\omega_n t} & x_c - \frac{L}{2} < x < x_c + \frac{L}{2} \\ 0 & \text{elsewhere} \end{cases}$$

where $k_n = \frac{n\pi}{L}$.

The unknown constant A may be found by normalizing the wave function, so that the total probability density of finding the particle in the system is 1. It follows that:

$$|A| = \sqrt{\frac{2}{L}}$$

Thus, A may be any complex number with an absolute value $\sqrt{2/L}$; these different values of A yield the same physical state, so $A = \sqrt{2/L}$ can be selected to simplify.

5.2 The position wave function probability distribution and *CPP*

In classical physics, the particle can be detected anywhere in the box with equal probability. In quantum mechanics, however, the probability density for finding a particle at a given position is derived from the wave function as $f(x) = |\psi(x)|^2$. For the particle in a box, the wave function position probability density function (*PDF*) for finding the particle at a given position depends upon its state and is given by:

$$f(x) = |\psi(x)|^2 = \begin{cases} \frac{2}{L}\sin^2\left[k_n\left(x - x_c + \frac{L}{2}\right)\right] & x_c - \frac{L}{2} < x < x_c + \frac{L}{2} \\ 0 & \text{otherwise} \end{cases}$$

Thus, for any value of n greater than one, there are regions within the box for which $f(x) = 0$, indicating that *spatial nodes* exist at which the particle cannot be found.

Therefore, the wave function position cumulative probability distribution function (*CDF*), which is equal to $P_r(X)$ in $\mathcal{R}$ is:

$$P_r(X) = F(x_j) = P_{rob}(X \le x_j) = \int_{-\infty}^{x_j} |\psi(x)|^2 dx$$

$$= \begin{cases} \int_{x_c - \frac{L}{2}}^{x_j} \frac{2}{L}\sin^2\left[k_n\left(x - x_c + \frac{L}{2}\right)\right]dx & x_c - \frac{L}{2} < x_j < x_c + \frac{L}{2} \\ 0 & \text{otherwise} \end{cases}$$

And the real complementary probability to $P_r(X)$ in $\mathcal{R}$, which is $P_m(X)/i$ is:

$$P_m(X)/i = 1 - P_r(X) = 1 - F(x_j) = 1 - P_{rob}(X \le x_j) = P_{rob}(X > x_j)$$
$$= 1 - \int_{-\infty}^{x_j} |\psi(x)|^2 dx = \int_{x_j}^{+\infty} |\psi(x)|^2 dx$$

$$= \begin{cases} 1 - \displaystyle\int_{x_c - \frac{L}{2}}^{x_j} \frac{2}{L} \sin^2\left[k_n\left(x - x_c + \frac{L}{2}\right)\right] dx & x_c - \frac{L}{2} < x_j < x_c + \frac{L}{2} \\ 0 & \text{otherwise} \end{cases}$$

$$= \begin{cases} \displaystyle\int_{x_j}^{x_c + \frac{L}{2}} \frac{2}{L} \sin^2\left[k_n\left(x - x_c + \frac{L}{2}\right)\right] dx & x_c - \frac{L}{2} < x_j < x_c + \frac{L}{2} \\ 0 & \text{otherwise} \end{cases}$$

Consequently, the imaginary complementary probability to $P_r(X)$ in $\mathcal{M}$, which is $P_m(X)$ is:

$$P_m(X) = i[1 - P_r(X)] = i[1 - F(x_j)] = i[1 - P_{rob}(X \le x_j)] = iP_{rob}(X > x_j)$$
$$= i\left[1 - \int_{-\infty}^{x_j} |\psi(x)|^2 dx\right] = i\int_{x_j}^{+\infty} |\psi(x)|^2 dx$$

$$= \begin{cases} i\left[1 - \displaystyle\int_{x_c - \frac{L}{2}}^{x_j} \frac{2}{L} \sin^2\left[k_n\left(x - x_c + \frac{L}{2}\right)\right] dx\right] & x_c - \frac{L}{2} < x_j < x_c + \frac{L}{2} \\ 0 & \text{otherwise} \end{cases}$$

$$= \begin{cases} i\left[\displaystyle\int_{x_j}^{x_c + \frac{L}{2}} \frac{2}{L} \sin^2\left[k_n\left(x - x_c + \frac{L}{2}\right)\right] dx\right] & x_c - \frac{L}{2} < x_j < x_c + \frac{L}{2} \\ 0 & \text{otherwise} \end{cases}$$

Furthermore, the complex random number or vector in $\mathcal{C} = \mathcal{R} + \mathcal{M}$, which is $Z(X)$ is:

$$
\begin{aligned}
Z(X) &= P_r(X) + P_m(X) = P_r(X) + i[1 - P_r(X)] = F(x_j) + i[1 - F(x_j)] \\
&= P_{rob}(X \le x_j) + i[1 - P_{rob}(X \le x_j)] = P_{rob}(X \le x_j) + iP_{rob}(X > x_j) \\
&= \int_{-\infty}^{x_j} |\psi(x)|^2 dx + i\left[1 - \int_{-\infty}^{x_j} |\psi(x)|^2 dx\right] = \int_{-\infty}^{x_j} |\psi(x)|^2 dx + i\int_{x_j}^{+\infty} |\psi(x)|^2 dx \\
&= \begin{cases} \displaystyle\int_{x_c-\frac{L}{2}}^{x_j} \frac{2}{L}\sin^2\left[k_n\left(x - x_c + \frac{L}{2}\right)\right]dx + i\left[1 - \int_{x_c-\frac{L}{2}}^{x_j} \frac{2}{L}\sin^2\left[k_n\left(x - x_c + \frac{L}{2}\right)\right]dx\right] & x_c - \frac{L}{2} < x_j < x_c + \frac{L}{2} \\ 0 & \text{otherwise} \end{cases} \\
&= \begin{cases} \displaystyle\int_{x_c-\frac{L}{2}}^{x_j} \frac{2}{L}\sin^2\left[k_n\left(x - x_c + \frac{L}{2}\right)\right]dx + i\left[\int_{x_j}^{x_c+\frac{L}{2}} \frac{2}{L}\sin^2\left[k_n\left(x - x_c + \frac{L}{2}\right)\right]dx\right] & x_c - \frac{L}{2} < x_j < x_c + \frac{L}{2} \\ 0 & \text{otherwise} \end{cases}
\end{aligned}
$$

Additionally, the degree of our knowledge, which is $DOK(X)$ is:

$$
\begin{aligned}
DOK(X) &= [P_r(X)]^2 + [P_m(X)/i]^2 = [P_r(X)]^2 + [1 - P_r(X)]^2 \\
&= [F(x_j)]^2 + [1 - F(x_j)]^2 = [P_{rob}(X \le x_j)]^2 + [1 - P_{rob}(X \le x_j)]^2 \\
&= [P_{rob}(X \le x_j)]^2 + [P_{rob}(X > x_j)]^2 \\
&= \left[\int_{-\infty}^{x_j} |\psi(x)|^2 dx\right]^2 + \left[1 - \int_{-\infty}^{x_j} |\psi(x)|^2 dx\right]^2 \\
&= \left[\int_{-\infty}^{x_j} |\psi(x)|^2 dx\right]^2 + \left[\int_{x_j}^{+\infty} |\psi(x)|^2 dx\right]^2 \\
&= \begin{cases} \displaystyle\left[\int_{x_c-\frac{L}{2}}^{x_j} \frac{2}{L}\sin^2\left[k_n\left(x - x_c + \frac{L}{2}\right)\right]dx\right]^2 + \left[1 - \int_{x_c-\frac{L}{2}}^{x_j} \frac{2}{L}\sin^2\left[k_n\left(x - x_c + \frac{L}{2}\right)\right]dx\right]^2 & x_c - \frac{L}{2} < x_j < x_c + \frac{L}{2} \\ 0 & \text{otherwise} \end{cases} \\
&= \begin{cases} \displaystyle\left[\int_{x_c-\frac{L}{2}}^{x_j} \frac{2}{L}\sin^2\left[k_n\left(x - x_c + \frac{L}{2}\right)\right]dx\right]^2 + \left[\int_{x_j}^{x_c+\frac{L}{2}} \frac{2}{L}\sin^2\left[k_n\left(x - x_c + \frac{L}{2}\right)\right]dx\right]^2 & x_c - \frac{L}{2} < x_j < x_c + \frac{L}{2} \\ 0 & \text{otherwise} \end{cases}
\end{aligned}
$$

Moreover, the chaotic factor, which is $Chf(X)$ is:

$$\begin{aligned}
Chf(X) &= 2iP_r(X)P_m(X) \\
&= 2iP_r(X) \times i[1 - P_r(X)] = -2P_r(X)[1 - P_r(X)] = -2F(x_j)[1 - F(x_j)] \\
&= -2P_{rob}(X \le x_j)[1 - P_{rob}(X \le x_j)] = -2P_{rob}(X \le x_j)P_{rob}(X > x_j) \\
&= -2\int_{-\infty}^{x_j} |\psi(x)|^2 dx \times \left[1 - \int_{-\infty}^{x_j} |\psi(x)|^2 dx\right] \\
&= -2\int_{-\infty}^{x_j} |\psi(x)|^2 dx \times \int_{x_j}^{+\infty} |\psi(x)|^2 dx \\
&= \begin{cases} -2\displaystyle\int_{x_c - \frac{L}{2}}^{x_j} \frac{2}{L}\sin^2\left[k_n\left(x - x_c + \frac{L}{2}\right)\right]dx \times \left[1 - \displaystyle\int_{x_c - \frac{L}{2}}^{x_j} \frac{2}{L}\sin^2\left[k_n\left(x - x_c + \frac{L}{2}\right)\right]dx\right] & x_c - \frac{L}{2} < x_j < x_c + \frac{L}{2} \\ 0 & \text{otherwise} \end{cases} \\
&= \begin{cases} -2\displaystyle\int_{x_c - \frac{L}{2}}^{x_j} \frac{2}{L}\sin^2\left[k_n\left(x - x_c + \frac{L}{2}\right)\right]dx \times \displaystyle\int_{x_j}^{x_c + \frac{L}{2}} \frac{2}{L}\sin^2\left[k_n\left(x - x_c + \frac{L}{2}\right)\right]dx & x_c - \frac{L}{2} < x_j < x_c + \frac{L}{2} \\ 0 & \text{otherwise} \end{cases}
\end{aligned}$$

In addition, the magnitude of the chaotic factor, which is $MChf(X)$ is:

$$\begin{aligned}
MChf(X) &= |Chf(X)| = -2iP_r(X)P_m(X) = -2iP_r(X) \times i[1 - P_r(X)] \\
&= 2P_r(X)[1 - P_r(X)] = 2F(x_j)[1 - F(x_j)] \\
&= 2P_{rob}(X \le x_j)[1 - P_{rob}(X \le x_j)] = 2P_{rob}(X \le x_j)P_{rob}(X > x_j) \\
&= 2\int_{-\infty}^{x_j} |\psi(x)|^2 dx \times \left[1 - \int_{-\infty}^{x_j} |\psi(x)|^2 dx\right] = 2\int_{-\infty}^{x_j} |\psi(x)|^2 dx \times \int_{x_j}^{+\infty} |\psi(x)|^2 dx \\
&= \begin{cases} 2\displaystyle\int_{x_c - \frac{L}{2}}^{x_j} \frac{2}{L}\sin^2\left[k_n\left(x - x_c + \frac{L}{2}\right)\right]dx \times \left[1 - \displaystyle\int_{x_c - \frac{L}{2}}^{x_j} \frac{2}{L}\sin^2\left[k_n\left(x - x_c + \frac{L}{2}\right)\right]dx\right] & x_c - \frac{L}{2} < x_j < x_c + \frac{L}{2} \\ 0 & \text{otherwise} \end{cases} \\
&= \begin{cases} 2\displaystyle\int_{x_c - \frac{L}{2}}^{x_j} \frac{2}{L}\sin^2\left[k_n\left(x - x_c + \frac{L}{2}\right)\right]dx \times \displaystyle\int_{x_j}^{x_c + \frac{L}{2}} \frac{2}{L}\sin^2\left[k_n\left(x - x_c + \frac{L}{2}\right)\right]dx & x_c - \frac{L}{2} < x_j < x_c + \frac{L}{2} \\ 0 & \text{otherwise} \end{cases}
\end{aligned}$$

Finally, the real probability in the complex probability universe $\mathcal{C} = \mathcal{R} + \mathcal{M}$ which is $Pc(X)$ is:

$$\begin{aligned}
Pc^2(X) &= \{[P_r(X)] + [P_m(X)/i]\}^2 = \{[P_r(X)] + [1 - P_r(X)]\}^2 = \left\{[F(x_j)] + [1 - F(x_j)]\right\}^2 \\
&= \left\{P_{rob}(X \le x_j) + [1 - P_{rob}(X \le x_j)]\right\}^2 = \left\{P_{rob}(X \le x_j) + P_{rob}(X > x_j)\right\}^2 \\
&= \left\{\int_{-\infty}^{x_j} |\psi(x)|^2 dx + \left[1 - \int_{-\infty}^{x_j} |\psi(x)|^2 dx\right]\right\}^2 = \left\{\int_{-\infty}^{x_j} |\psi(x)|^2 dx + \int_{x_j}^{+\infty} |\psi(x)|^2 dx\right\}^2 = \left\{\int_{-\infty}^{+\infty} |\psi(x)|^2 dx\right\}^2 \\
&= \begin{cases} \left\{\int_{x_c - \frac{L}{2}}^{x_j} \frac{2}{L}\sin^2\left[k_n\left(x - x_c + \frac{L}{2}\right)\right]dx + \left[1 - \int_{x_c - \frac{L}{2}}^{x_j} \frac{2}{L}\sin^2\left[k_n\left(x - x_c + \frac{L}{2}\right)\right]dx\right]\right\}^2 & x_c - \frac{L}{2} < x_j < x_c + \frac{L}{2} \\ 0 & \text{otherwise} \end{cases} \\
&= \begin{cases} \left\{\int_{x_c - \frac{L}{2}}^{x_j} \frac{2}{L}\sin^2\left[k_n\left(x - x_c + \frac{L}{2}\right)\right]dx + \int_{x_j}^{x_c + \frac{L}{2}} \frac{2}{L}\sin^2\left[k_n\left(x - x_c + \frac{L}{2}\right)\right]dx\right\}^2 & x_c - \frac{L}{2} < x_j < x_c + \frac{L}{2} \\ 0 & \text{otherwise} \end{cases} \\
&= \begin{cases} \left\{\int_{x_c - \frac{L}{2}}^{x_c + \frac{L}{2}} \frac{2}{L}\sin^2\left[k_n\left(x - x_c + \frac{L}{2}\right)\right]dx\right\}^2 & x_c - \frac{L}{2} < x_j < x_c + \frac{L}{2} \\ 0 & \text{otherwise} \end{cases} \\
&= \begin{cases} 1^2 & x_c - \frac{L}{2} < x_j < x_c + \frac{L}{2} \\ 0 & \text{otherwise} \end{cases} = \begin{cases} 1 & x_c - \frac{L}{2} < x_j < x_c + \frac{L}{2} \\ 0 & \text{otherwise} \end{cases} \\
&= Pc(X)
\end{aligned}$$

And, $Pc(X)$ can be computed using *CPP* as follows:

$$\begin{aligned}
Pc^2(X) &= DOK(X) - Chf(X) = [P_r(X)]^2 + [P_m(X)/i]^2 - 2iP_r(X)P_m(X) \\
&= [P_r(X)]^2 + [1 - P_r(X)]^2 + 2P_r(X)[1 - P_r(X)] = \{P_r(X) + [1 - P_r(X)]\}^2 \\
&= \left\{\int_{-\infty}^{x_j} |\psi(x)|^2 dx + \left[1 - \int_{-\infty}^{x_j} |\psi(x)|^2 dx\right]\right\}^2 = \left\{\int_{-\infty}^{x_j} |\psi(x)|^2 dx + \int_{x_j}^{+\infty} |\psi(x)|^2 dx\right\}^2 = \left\{\int_{-\infty}^{+\infty} |\psi(x)|^2 dx\right\}^2 \\
&= \begin{cases} 1^2 & x_c - \frac{L}{2} < x_j < x_c + \frac{L}{2} \\ 0 & \text{otherwise} \end{cases} = \begin{cases} 1 & x_c - \frac{L}{2} < x_j < x_c + \frac{L}{2} \\ 0 & \text{otherwise} \end{cases} \\
&= Pc(X)
\end{aligned}$$

And, $Pc(X)$ can be computed using always *CPP* as follows also:

$$\begin{aligned}
Pc^2(X) &= DOK(X) + MChf(X) = [P_r(X)]^2 + [P_m(X)/i]^2 + [-2iP_r(X)P_m(X)] \\
&= [P_r(X)]^2 + [1 - P_r(X)]^2 + 2P_r(X)[1 - P_r(X)] = \{P_r(X) + [1 - P_r(X)]\}^2 \\
&= \left\{\int_{-\infty}^{x_j} |\psi(x)|^2 dx + \left[1 - \int_{-\infty}^{x_j} |\psi(x)|^2 dx\right]\right\}^2 \\
&= \left\{\int_{-\infty}^{x_j} |\psi(x)|^2 dx + \int_{x_j}^{+\infty} |\psi(x)|^2 dx\right\}^2 = \left\{\int_{-\infty}^{+\infty} |\psi(x)|^2 dx\right\}^2
\end{aligned}$$

$$= \begin{cases} 1^2 & x_c - \frac{L}{2} < x_j < x_c + \frac{L}{2} \\ 0 & \text{otherwise} \end{cases} = \begin{cases} 1 & x_c - \frac{L}{2} < x_j < x_c + \frac{L}{2} \\ 0 & \text{otherwise} \end{cases}$$
$$= Pc(X)$$

Hence, the prediction of all the wave function position probabilities of the random infinite potential well problem in the $\mathcal{C} = \mathcal{R} + \mathcal{M}$ is permanently certain and perfectly deterministic.

Now, if $x_c - \frac{L}{2} \le L_b$ (Lower bound of x_j), U_b (Upper bound of x_j) $\le x_c + \frac{L}{2}$

$$\int_{L_b}^{U_b} \frac{2}{L} \sin^2\left[k_n\left(x - x_c + \frac{L}{2}\right)\right] dx = \frac{2}{L} \int_{L_b}^{U_b} \left\{ \frac{1 - \cos\left[2k_n\left(x - x_c + \frac{L}{2}\right)\right]}{2} \right\} dx$$
$$= \frac{1}{L} \int_{L_b}^{U_b} \left\{ 1 - \cos\left[2k_n\left(x - x_c + \frac{L}{2}\right)\right] \right\} dx$$
$$= \frac{1}{L} \left\{ x - \frac{\sin\left[2k_n\left(x - x_c + \frac{L}{2}\right)\right]}{2k_n} \right\}_{L_b}^{U_b}$$
$$= \frac{1}{2k_n L} \left\{ \left[2k_n U_b - \sin\left[2k_n\left(U_b - x_c + \frac{L}{2}\right)\right]\right] - \left[2k_n L_b - \sin\left[2k_n\left(L_b - x_c + \frac{L}{2}\right)\right]\right] \right\}$$

Thus,

$$\int_{-\infty}^{+\infty} f(x)dx = \int_{-\infty}^{x_c - \frac{L}{2}} f(x)dx + \int_{x_c - \frac{L}{2}}^{x_c + \frac{L}{2}} f(x)dx + \int_{x_c + \frac{L}{2}}^{+\infty} f(x)dx$$
$$= 0 + \int_{L_b = x_c - \frac{L}{2}}^{U_b = x_c + \frac{L}{2}} |\psi(x)|^2 dx + 0 = \int_{L_b = x_c - \frac{L}{2}}^{U_b = x_c + \frac{L}{2}} \frac{2}{L} \sin^2\left[k_n\left(x - x_c + \frac{L}{2}\right)\right] dx$$
$$= \frac{1}{2k_n L} \left\{ \left[2k_n\left(x_c + \frac{L}{2}\right) - \sin\left[2k_n\left(x_c + \frac{L}{2} - x_c + \frac{L}{2}\right)\right]\right] - \left[2k_n\left(x_c - \frac{L}{2}\right) - \sin\left[2k_n\left(x_c - \frac{L}{2} - x_c + \frac{L}{2}\right)\right]\right] \right\}$$
$$= \frac{1}{2k_n L} \{[2k_n x_c + k_n L - \sin[2k_n L]] - [2k_n x_c - k_n L - \sin[2k_n(0)]]\}$$
$$= \frac{1}{2k_n L} \{2k_n L - \sin[2k_n L]\}$$

But $k_n = \frac{n\pi}{L}$, so it is equal to:

$$\frac{1}{2k_nL}\left\{2k_nL - \sin\left[\frac{2n\pi L}{L}\right]\right\} = \frac{1}{2k_nL}\{2k_nL - \sin[2n\pi]\} = \frac{1}{2k_nL}\{2k_nL - 0\},$$

where $n = 1,2,3,\ldots$

$$= \frac{2k_nL}{2k_nL} = 1$$

Therefore, $f(x) = |\psi(x)|^2$ is a probability density function since:

1. $\forall x : 0 \leq |\psi(x)|^2 \leq 1$, as $\forall x : -1 \leq \sin(x) \leq 1 \Leftrightarrow \forall x : 0 \leq \sin^2(x) \leq 1$

2. $\int_{-\infty}^{+\infty} |\psi(x)|^2 dx = 1$

Moreover, if $L_b = x_c - \frac{L}{2}$ and $x_c - \frac{L}{2} \leq (U_b = x_j) \leq x_c + \frac{L}{2}$, then:

$$\int_{x_c-\frac{L}{2}}^{x_j} \frac{2}{L}\sin^2\left[k_n\left(x - x_c + \frac{L}{2}\right)\right]dx$$

$$= \frac{1}{2k_nL}\left\{\left[2k_nx_j - \sin\left[2k_n\left(x_j - x_c + \frac{L}{2}\right)\right]\right] - \left[2k_n\left(x_c - \frac{L}{2}\right) - \sin\left[2k_n\left(x_c - \frac{L}{2} - x_c + \frac{L}{2}\right)\right]\right]\right\}$$

$$= \frac{1}{2k_nL}\left\{\left[2k_nx_j - \sin\left[2k_n\left(x_j - x_c + \frac{L}{2}\right)\right]\right] - [2k_nx_c - k_nL - \sin[2k_n(0)]]\right\}$$

$$= \frac{1}{2k_nL}\left\{\left[2k_nx_j - \sin\left[2k_n\left(x_j - x_c + \frac{L}{2}\right)\right]\right] - [2k_nx_c - k_nL]\right\}$$

$$= \frac{1}{2k_nL}\left\{2k_n\left(x_j - x_c + \frac{L}{2}\right) - \sin\left[2k_n\left(x_j - x_c + \frac{L}{2}\right)\right]\right\}$$

Additionally, if $x_c - \frac{L}{2} \leq (L_b = x_j) \leq x_c + \frac{L}{2}$ and $U_b = x_c + \frac{L}{2}$, then:

$$\int_{x_j}^{x_c+\frac{L}{2}} \frac{2}{L}\sin^2\left[k_n\left(x - x_c + \frac{L}{2}\right)\right]dx$$

$$= \frac{1}{2k_nL}\left\{\left[2k_n\left(x_c + \frac{L}{2}\right) - \sin\left[2k_n\left(x_c + \frac{L}{2} - x_c + \frac{L}{2}\right)\right]\right] - \left[2k_nx_j - \sin\left[2k_n\left(x_j - x_c + \frac{L}{2}\right)\right]\right]\right\}$$

$$= \frac{1}{2k_nL}\left\{[2k_nx_c + k_nL - \sin[2k_nL]] - \left[2k_nx_j - \sin\left[2k_n\left(x_j - x_c + \frac{L}{2}\right)\right]\right]\right\}$$

But $k_n = \frac{n\pi}{L}$

So, it is equal to:

$$\frac{1}{2k_nL}\left\{\left[2k_nx_c + k_nL - \sin\left[\frac{2n\pi L}{L}\right]\right] - \left[2k_nx_j - \sin\left[2k_n\left(x_j - x_c + \frac{L}{2}\right)\right]\right]\right\}$$

$$= \frac{1}{2k_nL}\left\{[2k_nx_c + k_nL - \sin[2n\pi]] - \left[2k_nx_j - \sin\left[2k_n\left(x_j - x_c + \frac{L}{2}\right)\right]\right]\right\},$$

where $n = 1,2,3, \ldots = \frac{1}{2k_nL}\left\{[2k_nx_c + k_nL - 0] - \left[2k_nx_j - \sin\left[2k_n\left(x_j - x_c + \frac{L}{2}\right)\right]\right]\right\}$

$$= \frac{1}{2k_nL}\left\{[2k_nx_c + k_nL] - \left[2k_nx_j - \sin\left[2k_n\left(x_j - x_c + \frac{L}{2}\right)\right]\right]\right\}$$

$$= \frac{1}{2k_nL}\left\{2k_n\left(x_c - x_j + \frac{L}{2}\right) + \sin\left[2k_n\left(x_j - x_c + \frac{L}{2}\right)\right]\right\}$$

5.3 The new model simulations

The following figures (**Figures 4–38**) illustrate all the calculations done above.

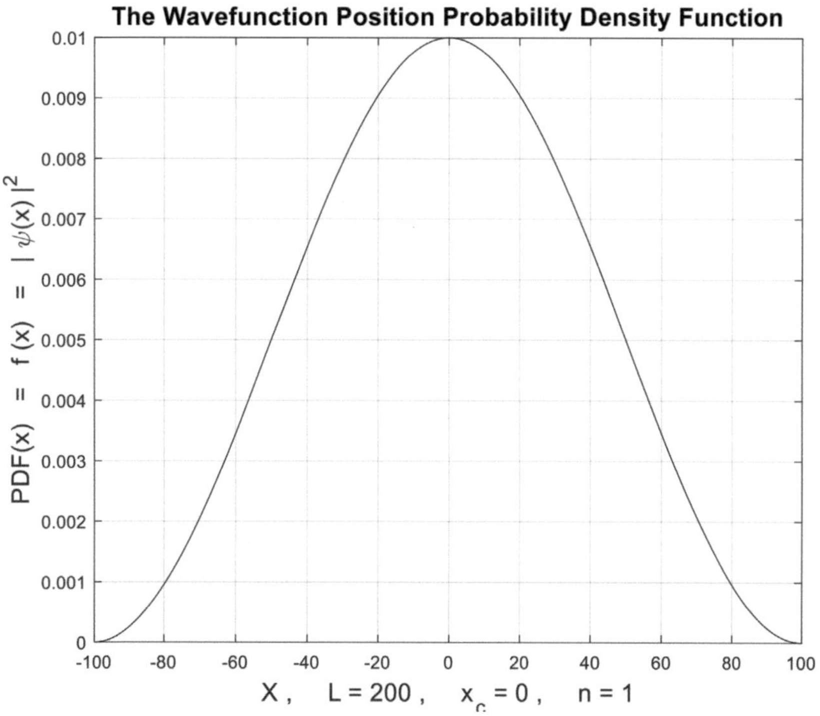

Figure 4.
The graph of the PDF *of the wave function position probability distribution as a function of the random variable* X *for* n = 1.

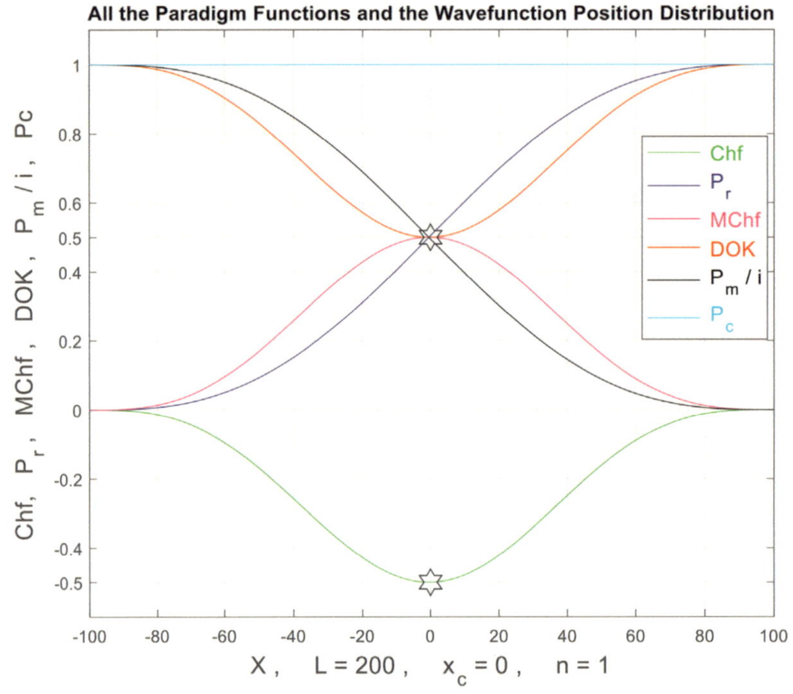

Figure 5.
The graphs of all the CPP *parameters as functions of the random variable* X *for the wave function position probability distribution for* n = 1.

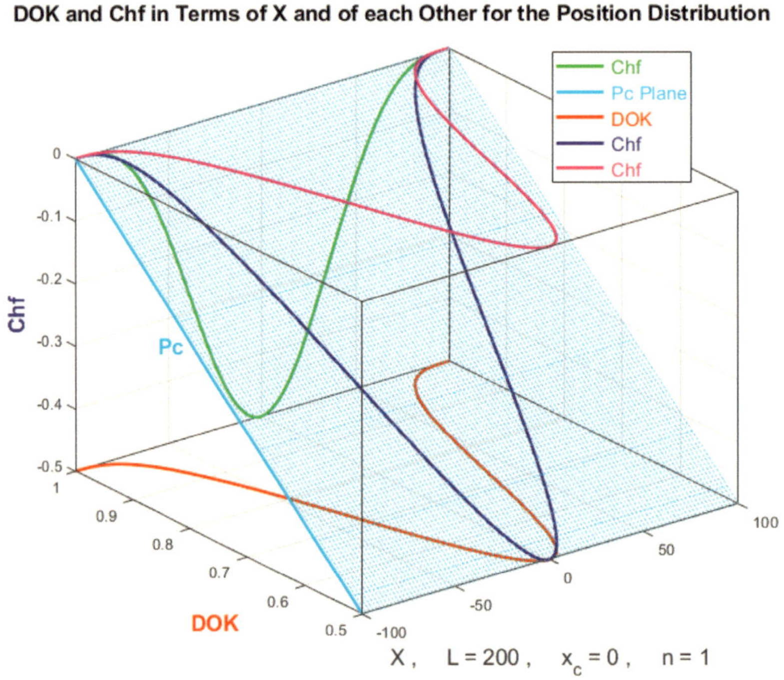

Figure 6.
The graphs of DOK *and* Chf,, *and the deterministic probability* Pc *in terms of* X *and of each other for the wave function position probability distribution for* n = 1.

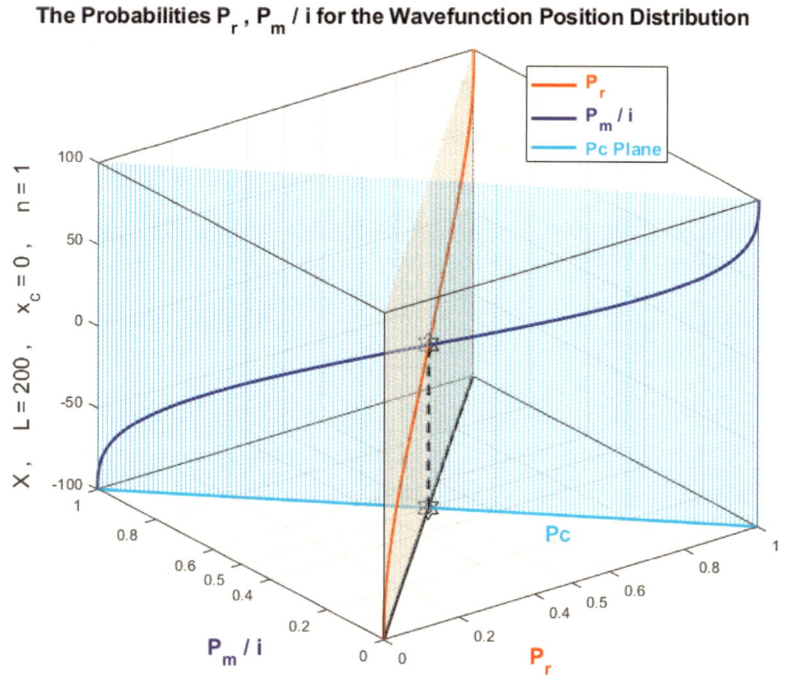

Figure 7.
The graphs of P_r *and* P_m/i, *and* Pc *in terms of* X *and of each other for the wave function position probability distribution for* n = 1.

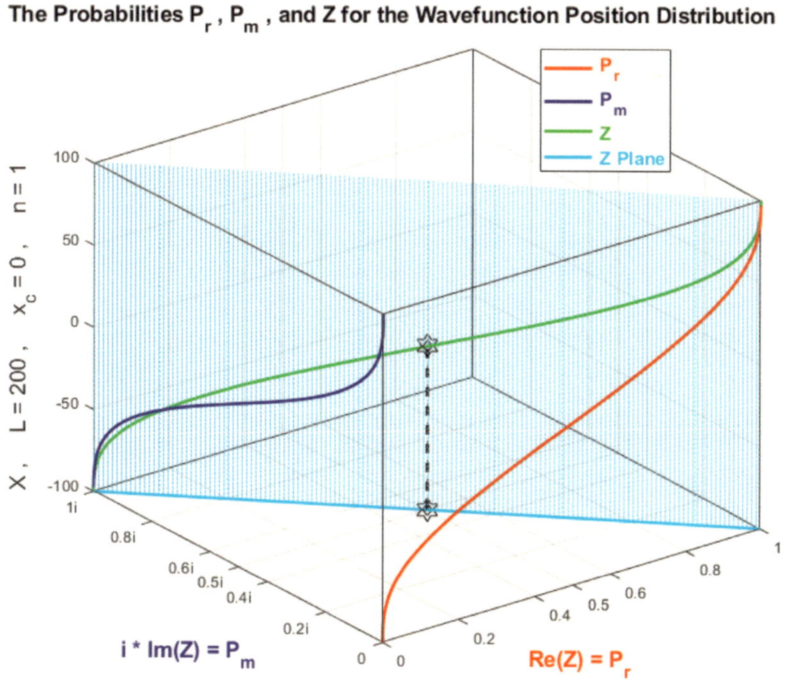

Figure 8.
The graphs of the probabilities P_r *and* P_m *and* Z *in terms of* X *for the wave function position probability distribution for* n = 1.

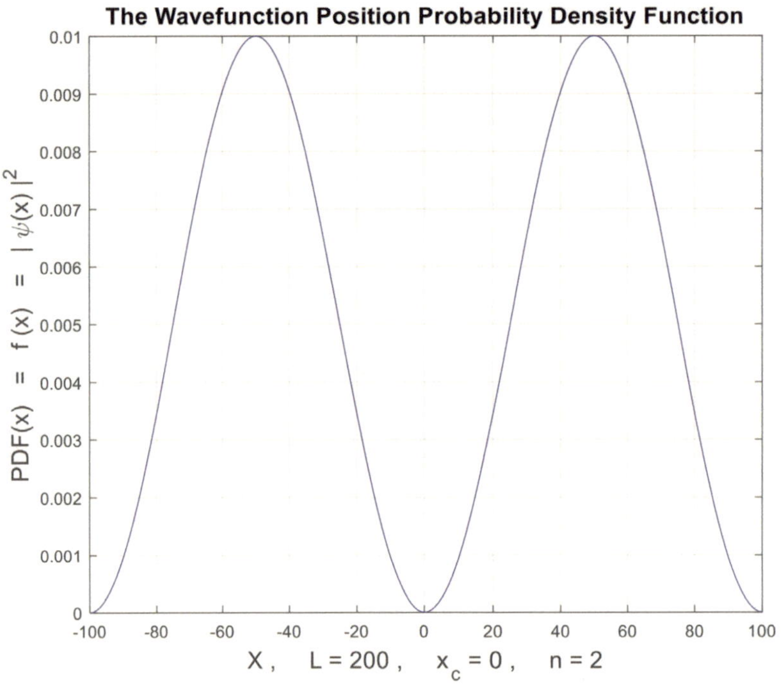

Figure 9.
The graph of the PDF *of the wave function position probability distribution as a function of the random variable* X *for* n = 2.

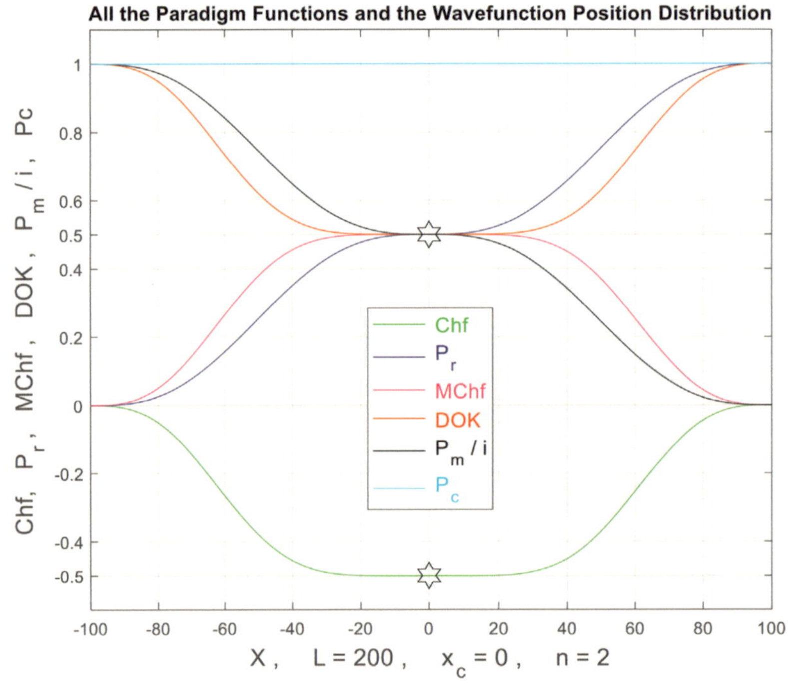

Figure 10.
The graphs of all the CPP *parameters as functions of the random variable* X *for the wave function position probability distribution for* n = 2.

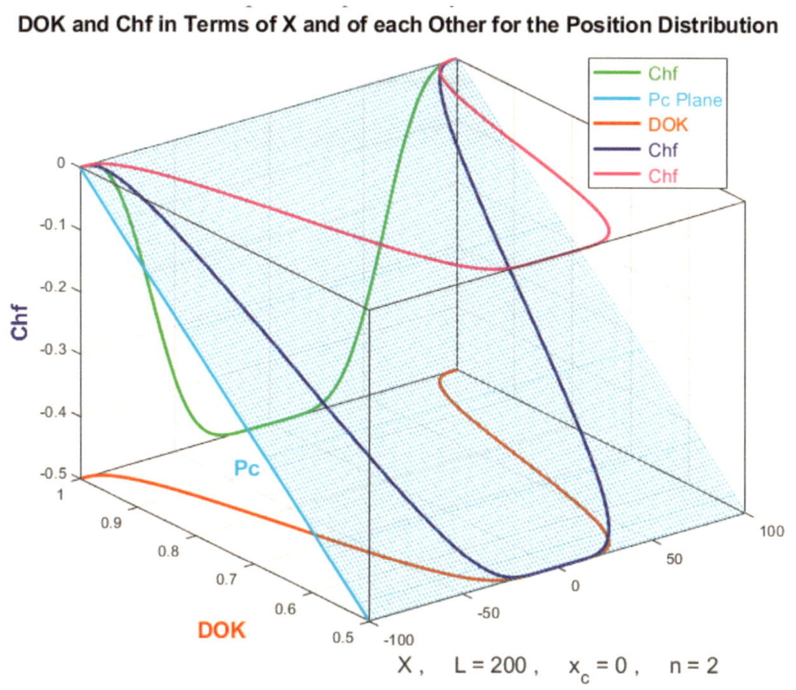

Figure 11.
The graphs of DOK *and* Chf, *and the deterministic probability* Pc *in terms of* X *and of each other for the wave function position probability distribution for* n = 2.

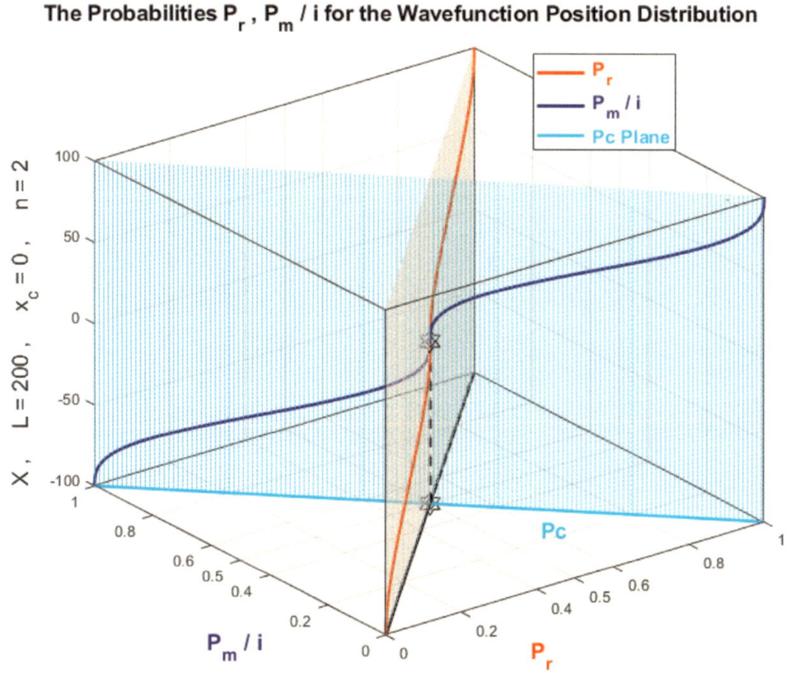

Figure 12.
The graphs of P_r *and* P_m/i, *and* Pc *in terms of* X *and of each other for the wave function position probability distribution for* n = 2.

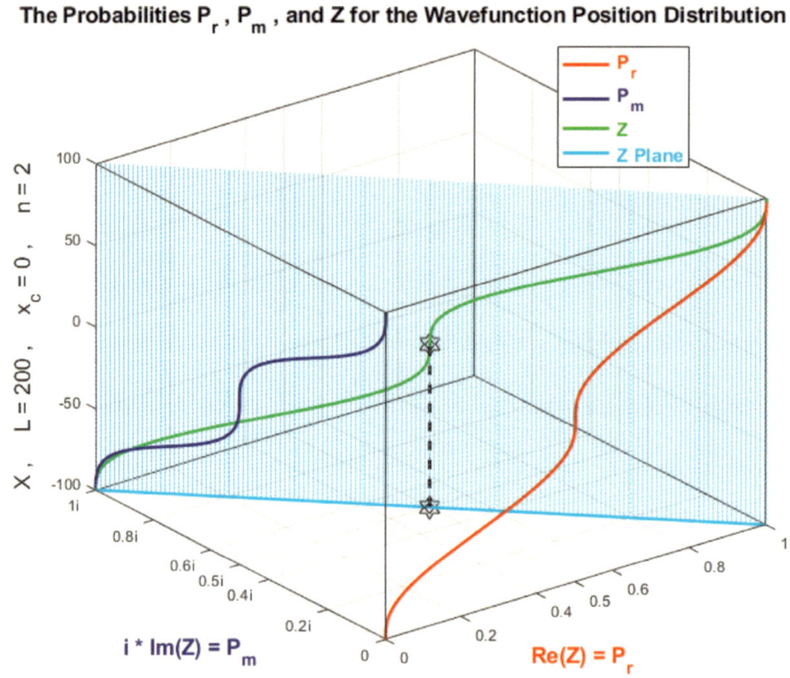

Figure 13.
The graphs of the probabilities P_r *and* P_m *and* Z *in terms of* X *for the wave function position probability distribution for* n = 2.

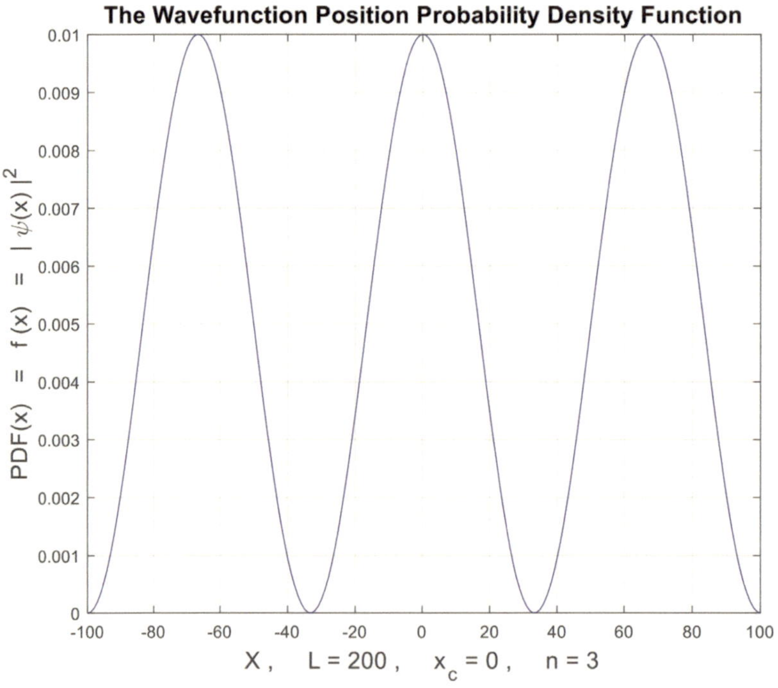

Figure 14.
The graph of the PDF *of the wave function position probability distribution as a function of the random variable* X *for* n = 3.

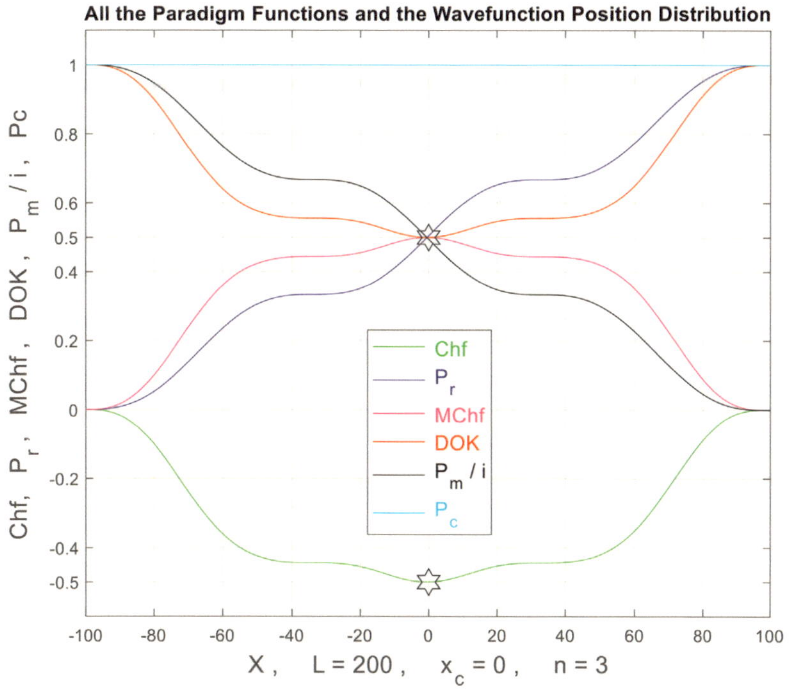

Figure 15.
The graphs of all the CPP *parameters as functions of the random variable* X *for the wave function position probability distribution for* n = *3.*

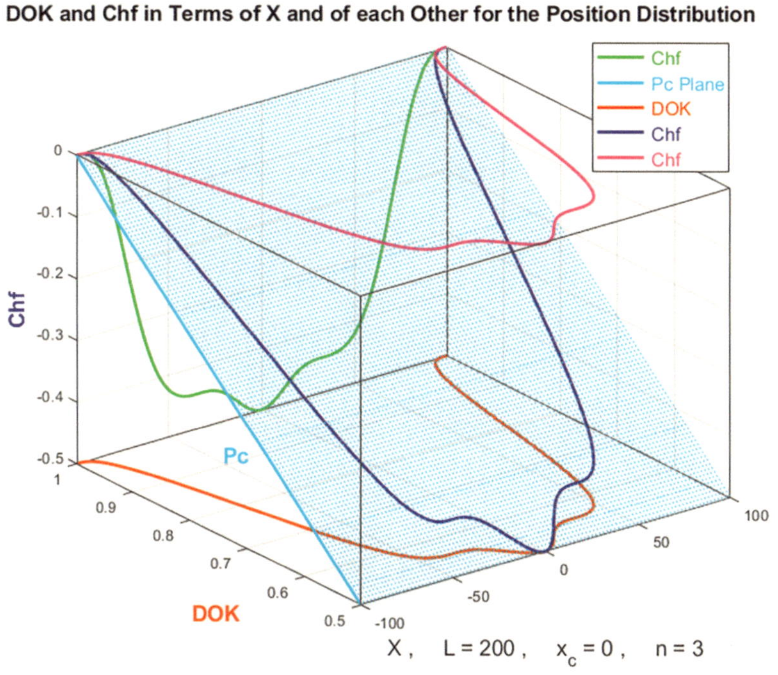

Figure 16.
The graphs of DOK *and* Chf, *and the deterministic probability* Pc *in terms of* X *and of each other for the wave function position probability distribution for* n = *3.*

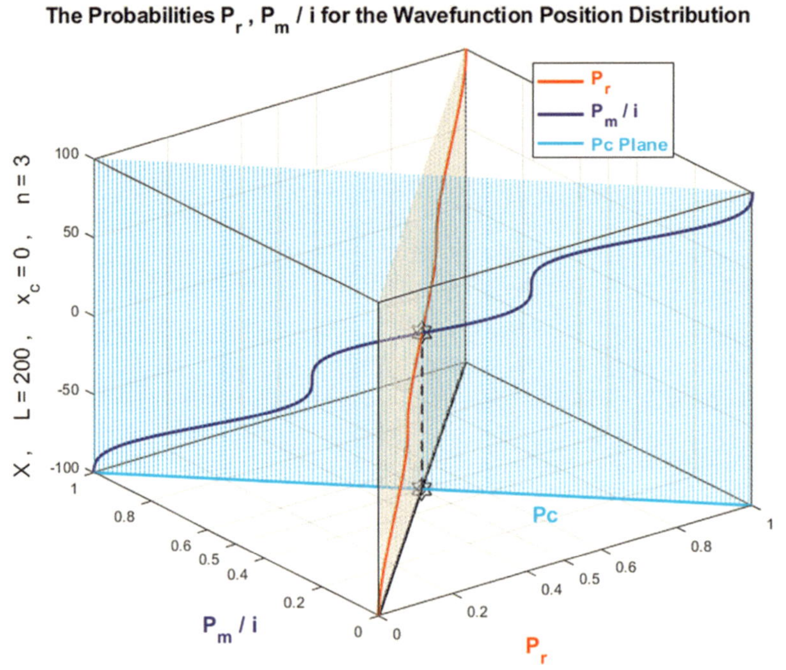

Figure 17.
The graphs of P_r *and* P_m/i, *and* Pc *in terms of* X *and of each other for the wave function position probability distribution for* n = 3.

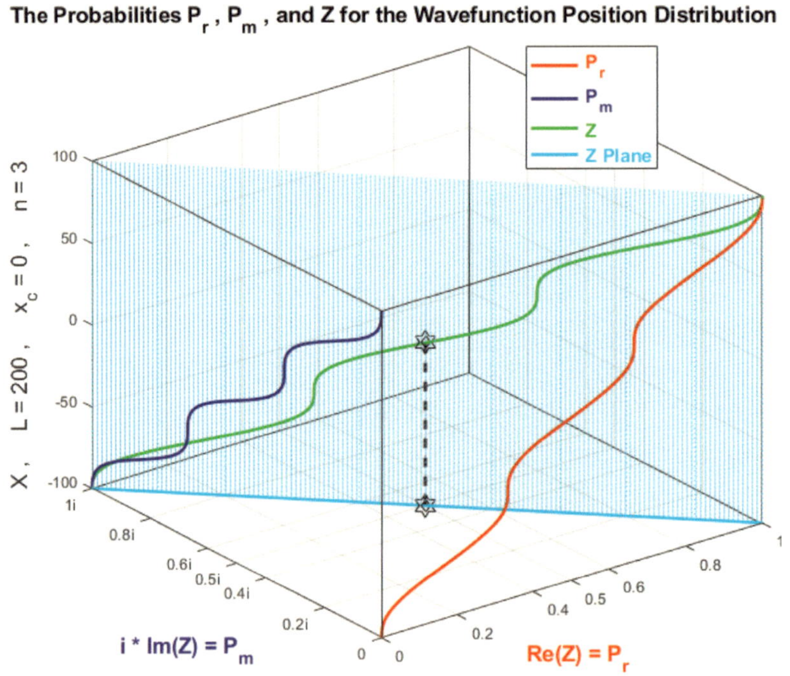

Figure 18.
The graphs of the probabilities P_r *and* P_m *and* Z *in terms of* X *for the wave function position probability distribution for* n = 3.

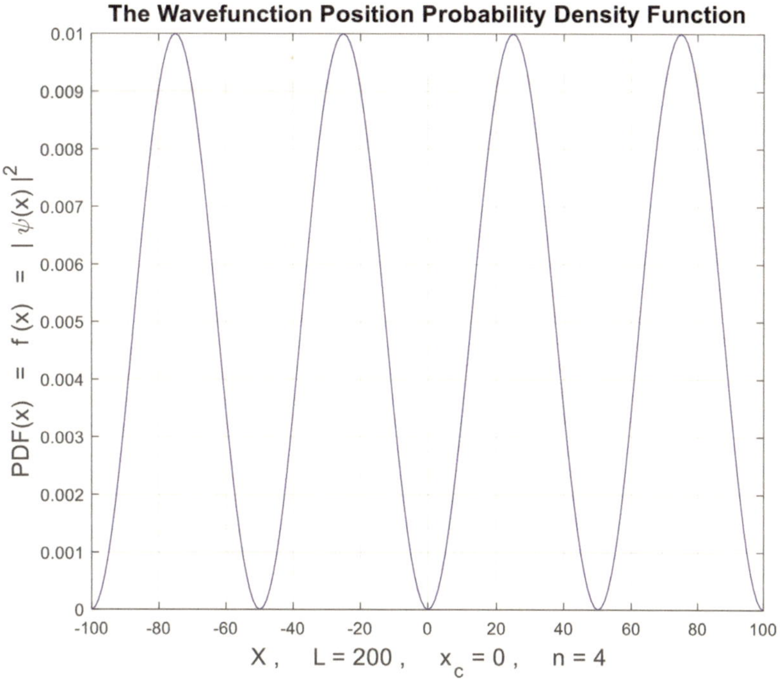

Figure 19.
The graph of the PDF *of the wave function position probability distribution as a function of the random variable* X *for* n = 4.

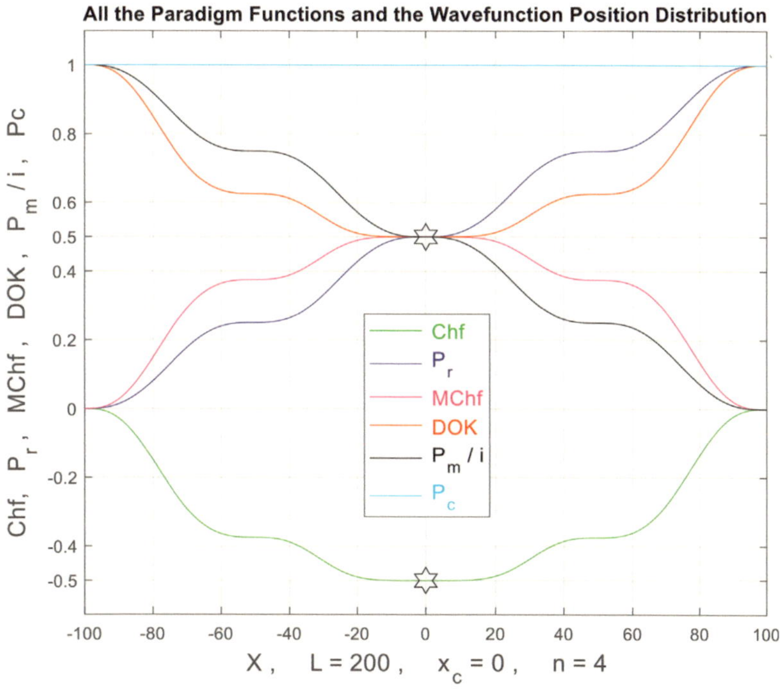

Figure 20.
The graphs of all the CPP *parameters as functions of the random variable* X *for the wave function position probability distribution for* n = 4.

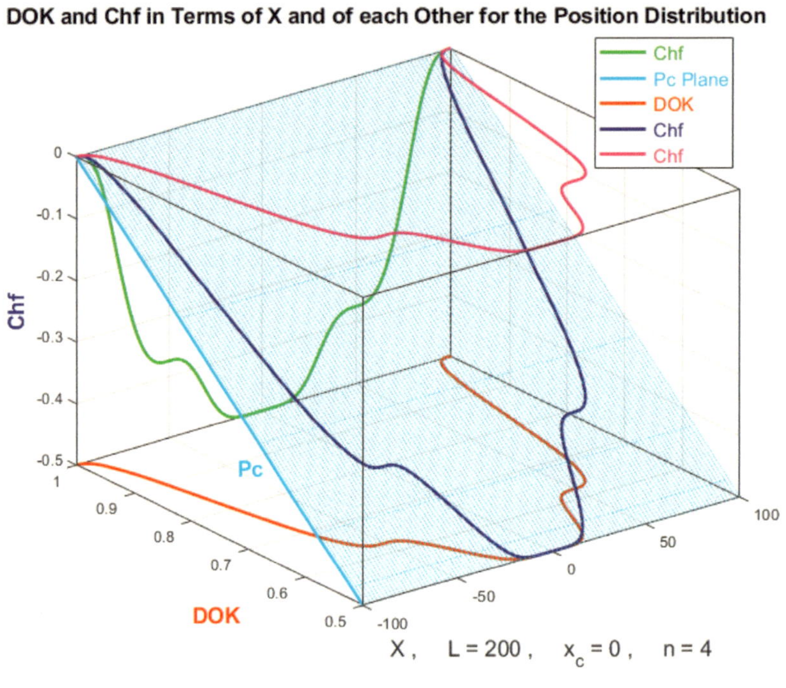

Figure 21.
The graphs of DOK *and* Chf, *and the deterministic probability* Pc *in terms of* X *and of each other for the wave function position probability distribution for* n = 4.

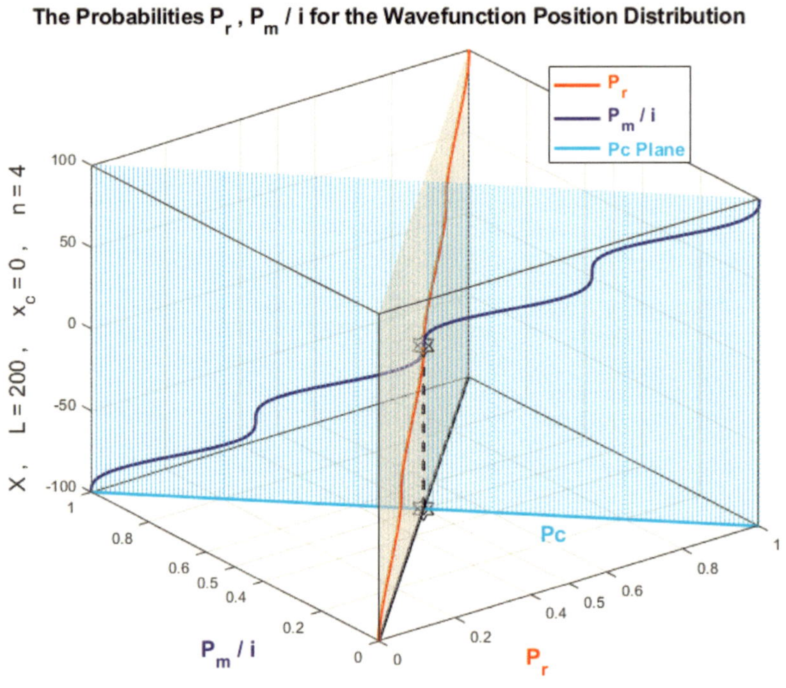

Figure 22.
The graphs of P_r *and* P_m/i, *and* Pc *in terms of* X *and of each other for the wave function position probability distribution for* n = 4.

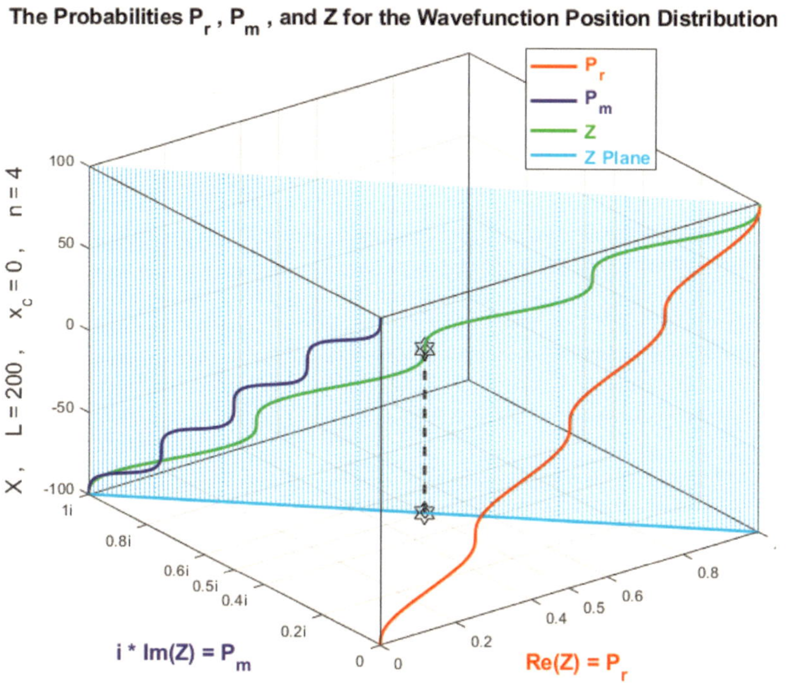

Figure 23.
The graphs of the probabilities P_r *and* P_m *and* Z *in terms of* X *for the wave function position probability distribution for* n = 4.

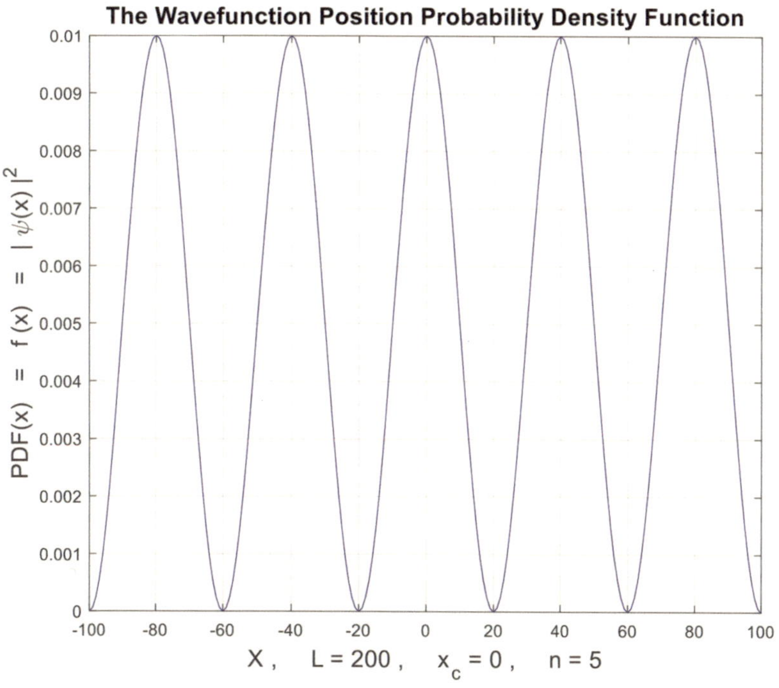

Figure 24.
The graph of the PDF *of the wave function position probability distribution as a function of the random variable* X *for* n = 5.

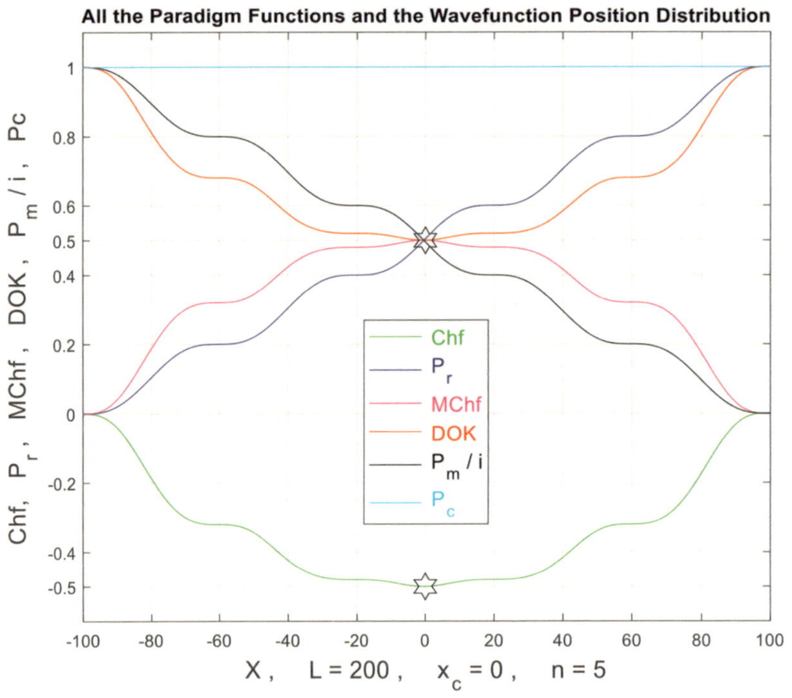

Figure 25.
The graphs of all the CPP *parameters as functions of the random variable* X *for the wave function position probability distribution for* n = 5.

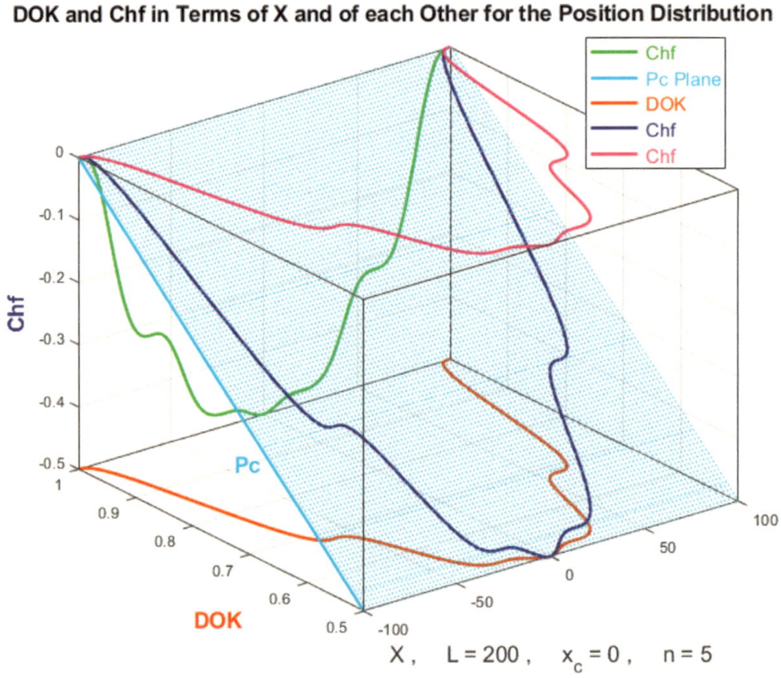

Figure 26.
The graphs of DOK *and* Chf, *and the deterministic probability* Pc *in terms of* X *and of each other for the wave function position probability distribution for* n = 5.

DOI: http://dx.doi.org/10.5772/intechopen.107300

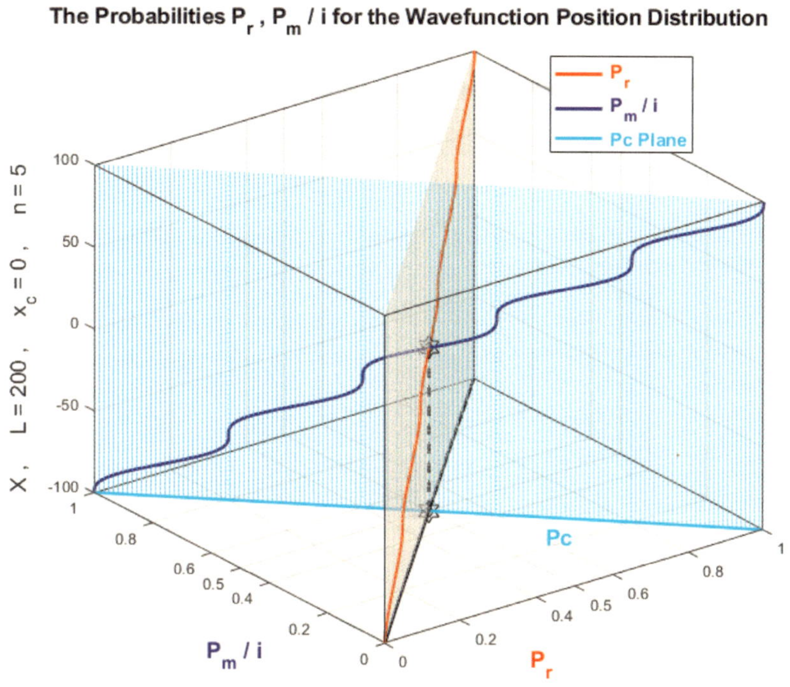

Figure 27.
The graphs of P_r *and* P_m/i, *and* Pc *in terms of* X *and of each other for the wave function position probability distribution for* $n = 5$.

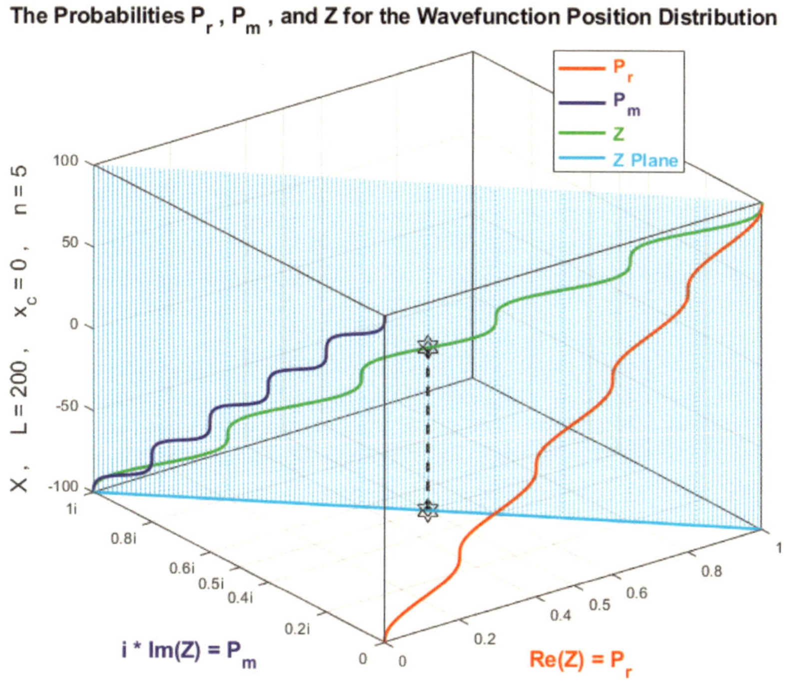

Figure 28.
The graphs of the probabilities P_r *and* P_m *and* Z *in terms of* X *for the wave function position probability distribution for* $n = 5$.

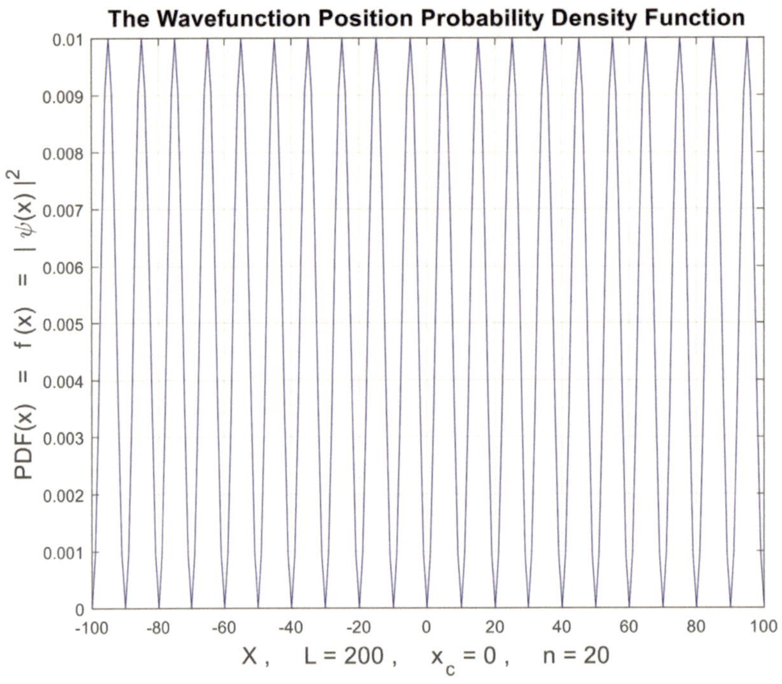

Figure 29.
The graph of the PDF *of the wave function position probability distribution as a function of the random variable* X *for* n = 20.

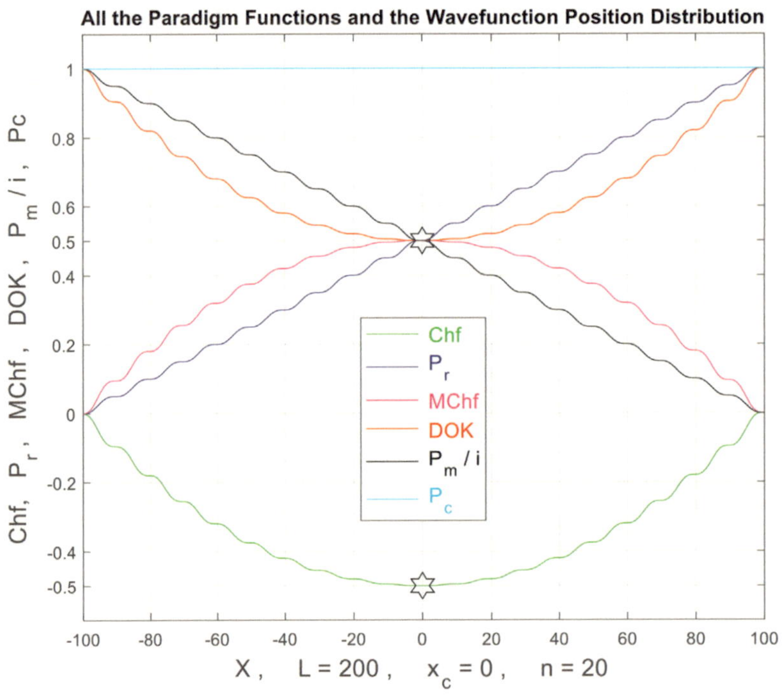

Figure 30.
The graphs of all the CPP *parameters as functions of the random variable* X *for the wave function position probability distribution for* n = 20.

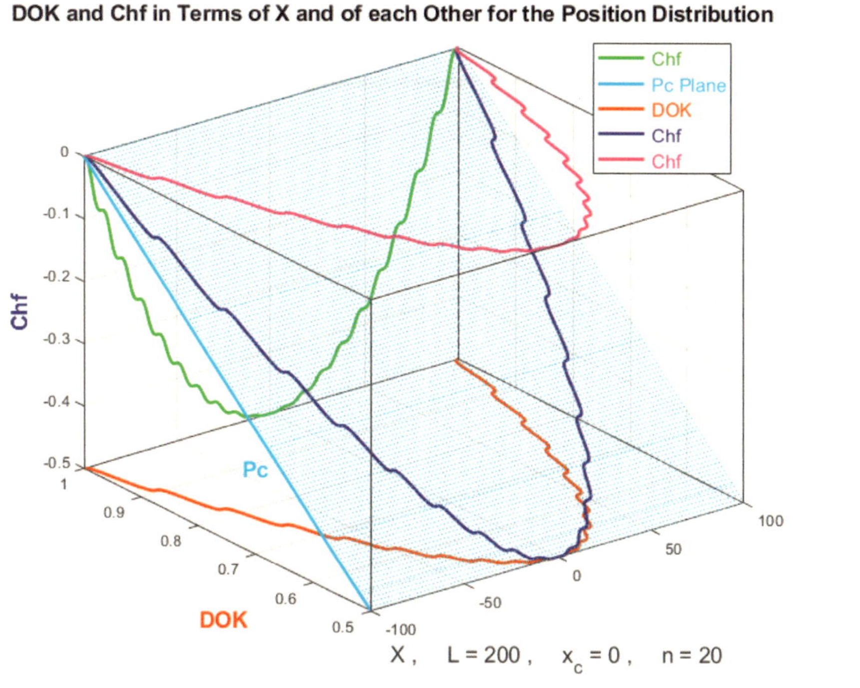

Figure 31.
The graphs of DOK *and* Chf, *and the deterministic probability* Pc *in terms of* X *and of each other for the wave function position probability distribution for* n = 20.

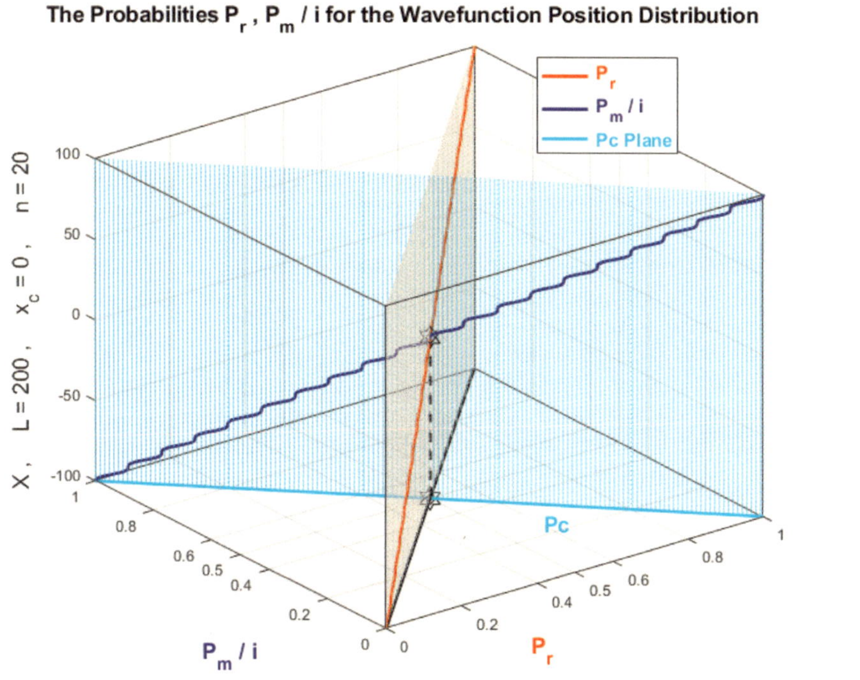

Figure 32.
The graphs of P_r *and* P_m/i, *and* Pc *in terms of* X *and of each other for the wave function position probability distribution for* n = 20.

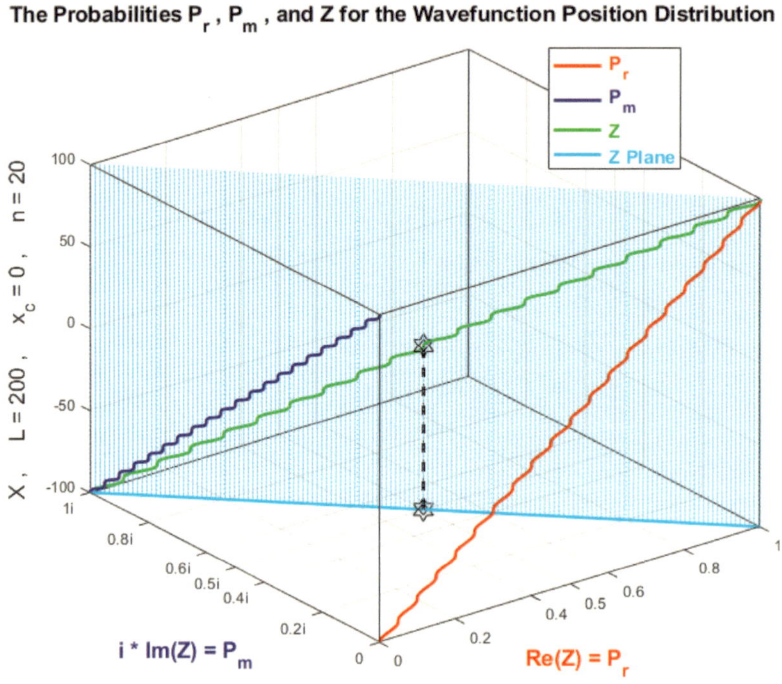

Figure 33.
The graphs of the probabilities P_r *and* P_m *and* Z *in terms of* X *for the wave function position probability distribution for* n = 20.

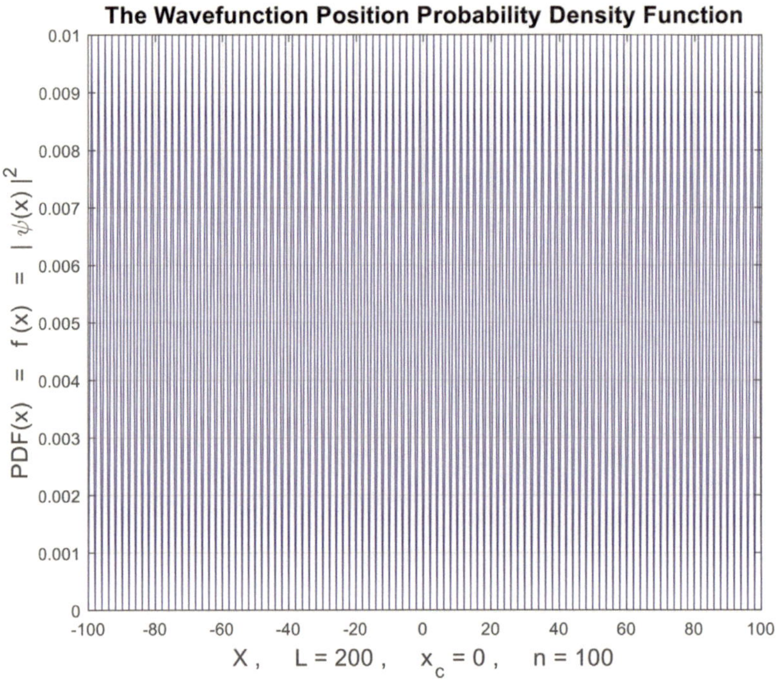

Figure 34.
The graph of the PDF *of the wave function position probability distribution as a function of the random variable* X *for* n = 100.

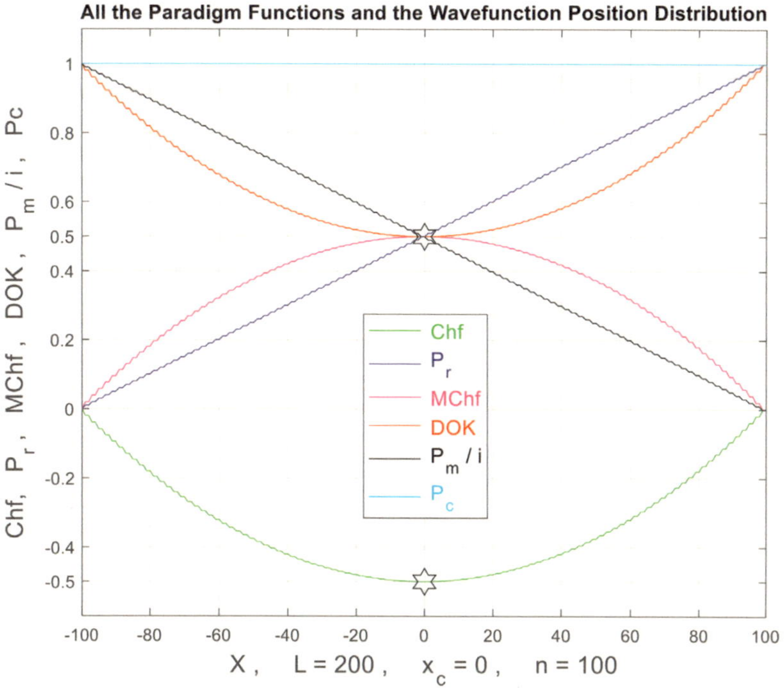

Figure 35.
The graphs of all the CPP *parameters as functions of the random variable* X *for the wave function position probability distribution for* n = 100.

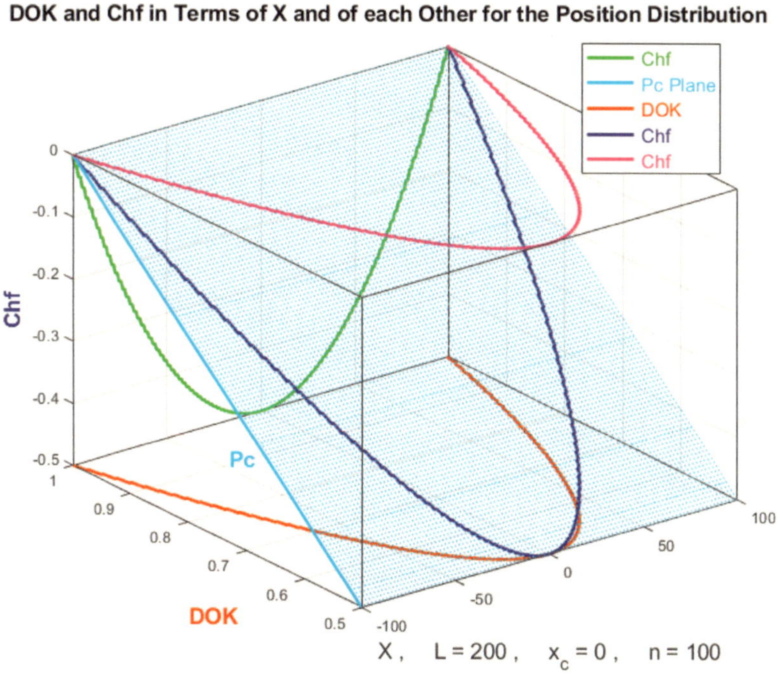

Figure 36.
The graphs of DOK *and* Chf, *and the deterministic probability* Pc *in terms of* X *and of each other for the wave function position probability distribution for* n = 100.

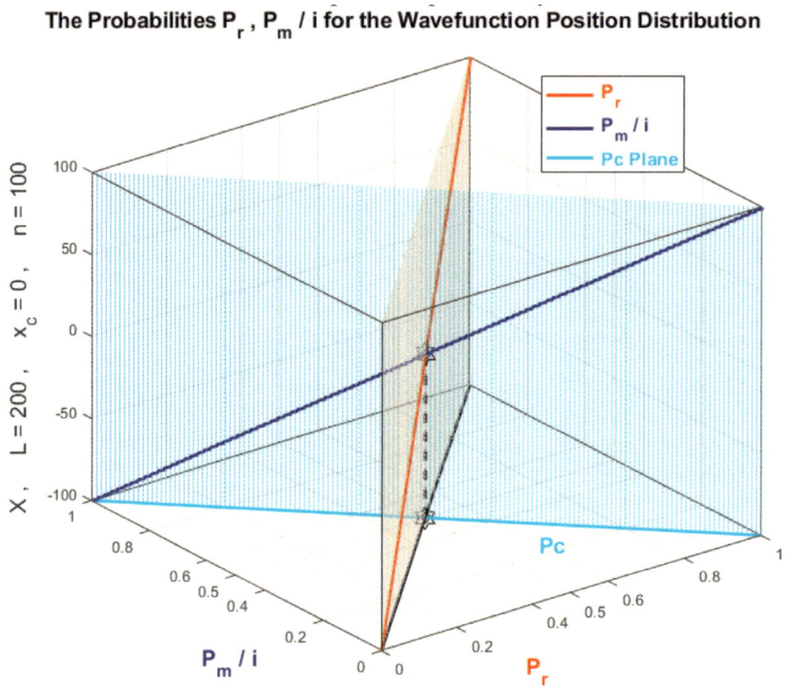

Figure 37.
The graphs of P_r *and* P_m/i, *and* Pc *in terms of* X *and of each other for the wave function position probability distribution for* n = 100.

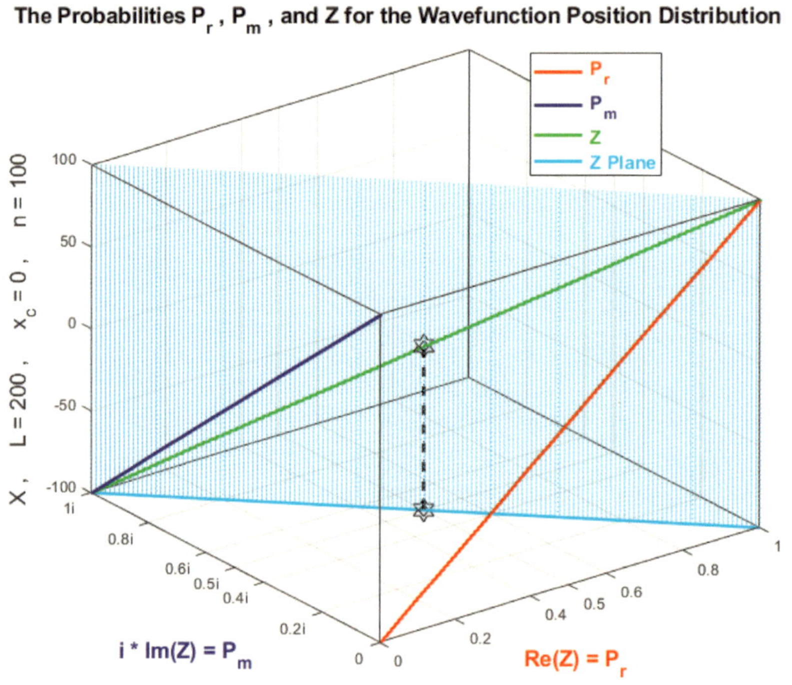

Figure 38.
The graphs of the probabilities P_r *and* P_m *and* Z *in terms of* X *for the wave function position probability distribution for* n = 100.

5.3.1 Simulations interpretation

In **Figures 4, 9, 14, 19, 24, 29,** and **34**, we can see the graphs of the probability density functions (*PDF*) of the wave function position probability distribution for this problem as functions of the random variable $X : -100 \leq X \leq 100$ for $n = 1, 2, 3, 4, 5, 20$, and 100.

In **Figures 5, 10, 15, 20, 25, 30,** and **35**, we can see also the graphs and the simulations of all the *CPP* parameters (*Chf*, *MChf*, *DOK*, P_r, P_m/i, and *Pc*) as functions of the random variable *X* for the wave function position probability distribution of the infinite potential well problem for $n = 1, 2, 3, 4, 5, 20$, and 100. Hence, we can visualize all the new paradigm functions for this problem.

In the cubes (**Figures 6, 11, 16, 21, 26, 31,** and **36**), the simulation of *DOK* and *Chf* as functions of each other and of the random variable *X* for the infinite potential well problem wave function position probability distribution can be seen. The thick line in cyan is the projection of the plane $Pc^2(X) = DOK(X) - Chf(X) = 1 = Pc(X)$ on the plane $X = L_b$ = lower bound of $X = -100$. This thick line starts at the point ($DOK = 1$, $Chf = 0$) when $X = L_b = -100$, reaches the point ($DOK = 0.5$, $Chf = -0.5$) when $X = 0$, and returns at the end to ($DOK = 1$, $Chf = 0$) when $X = U_b$ = upper bound of $X = 100$. The other curves are the graphs of $DOK(X)$ (red) and $Chf(X)$ (green, blue, and pink) in different simulation planes. Notice that they all have a minimum at the point ($DOK = 0.5$, $Chf = -0.5$, and $X = 0$). The last simulation point corresponds to ($DOK = 1$, $Chf = 0$, and $X = U_b = 100$).

In the cubes (**Figures 7, 12, 17, 22, 27, 32,** and **37**), we can notice the simulation of the real probability $P_r(X)$ in $\mathcal{R}$ and its complementary real probability $P_m(X)/i$ in $\mathcal{R}$ also in terms of the random variable *X* for the infinite potential well problem wave function position probability distribution. The thick line in cyan is the projection of the plane $Pc^2(X) = P_r(X) + P_m(X)/i = 1 = Pc(X)$ on the plane $X = L_b$ = lower bound of $X = -100$. This thick line starts at the point ($P_r = 0$, $P_m/i = 1$) and ends at the point ($P_r = 1$, $P_m/i = 0$). The red curve represents $P_r(X)$ in the plane $P_r(X) = P_m(X)/i$ in light gray. This curve starts at the point ($P_r = 0$, $P_m/i = 1$, and $X = L_b$ = lower bound of $X = -100$), reaches the point ($P_r = 0.5$, $P_m/i = 0.5$, and $X = 0$), and gets at the end to ($P_r = 1$, $P_m/i = 0$, and $X = U_b$ = upper bound of $X = 100$). The blue curve represents $P_m(X)/i$ in the plane in cyan $P_r(X) + P_m(X)/i = 1 = Pc(X)$. Notice the importance of the point, which is the intersection of the red and blue curves at $X = 0$, and when $P_r(X) = P_m(X)/i = 0.5$.

In the cubes (**Figures 8, 13, 18, 23, 28, 33,** and **38**), we can notice the simulation of the complex probability $Z(X)$ in $\mathcal{C} = \mathcal{R} + \mathcal{M}$ as a function of the real probability $P_r(X) = \text{Re}(Z)$ in $\mathcal{R}$ and of its complementary imaginary probability $P_m(X) = i \times \text{Im}(Z)$ in $\mathcal{M}$, and this in terms of the random variable *X* for the infinite potential well problem wave function position probability distribution. The red curve represents $P_r(X)$ in the plane $P_m(X) = 0$, and the blue curve represents $P_m(X)$ in the plane $P_r(X) = 0$. The green curve represents the complex probability $Z(X) = P_r(X) + P_m(X) = \text{Re}(Z) + i \times \text{Im}(Z)$ in the plane $P_r(X) = iP_m(X) + 1$ or $Z(X)$ plane in cyan. The curve of $Z(X)$ starts at the point ($P_r = 0$, $P_m = i$, and $X = L_b$ = lower bound of $X = -100$) and ends at the point ($P_r = 1$, $P_m = 0$, and $X = U_b$ = upper bound of $X = 100$). The thick line in cyan is $P_r(X = L_b = -100) = iP_m(X = L_b = -100) + 1$, and it is the projection of the $Z(X)$ curve on the complex probability plane whose equation is $X = L_b = -100$. This projected thick line starts at the point ($P_r = 0$, $P_m = i$, $X = L_b = -100$) and ends at the point ($P_r = 1$, $P_m = 0$, and $X = L_b = -100$). Notice the importance of the point corresponding to $X = 0$ and $Z = 0.5 + 0.5i$, when $P_r = 0.5$ and $P_m = 0.5i$.

5.4 The characteristics of the position probability distribution

In quantum mechanics, the average, or expectation value of the position of a particle is given by [10]:

$$\langle x\rangle = \int_{-\infty}^{+\infty} x|\psi(x)|^2 dx = \int_{x_c-\frac{L}{2}}^{x_c+\frac{L}{2}} \frac{2}{L} x \sin^2\left[k_n\left(x - x_c + \frac{L}{2}\right)\right] dx$$

For the steady state particle in a box, it can be shown that the average position is always $\langle x\rangle = x_c$, regardless of the state of the particle. For a superposition of states, the expectation value of the position will change based on the cross term, which is proportional to $\cos(\omega t)$. In the probability set and universe $\mathcal{R}$, we have:

$$\langle x\rangle_R = \langle x\rangle = x_c$$

The variance in the position is a measure of the uncertainty in the position of the particle, so in the probability set and universe $\mathcal{R}$, we have:

$$\mathrm{Var}_{x,R} = \mathrm{Var}(x) = \langle x^2\rangle_R - \langle x\rangle_R^2 = \left\{\int_{-\infty}^{+\infty} x^2|\psi(x)|^2 dx\right\} - \left\{\int_{-\infty}^{+\infty} x|\psi(x)|^2 dx\right\}^2$$

$$= \left\{\int_{x_c-\frac{L}{2}}^{x_c+\frac{L}{2}} \frac{2}{L} x^2 \sin^2\left[k_n\left(x - x_c + \frac{L}{2}\right)\right] dx\right\} - x_c^2 = \frac{L^2}{12}\left(1 - \frac{6}{n^2\pi^2}\right)$$

In the probability set and universe $\mathcal{M}$, we have:

$$\langle x\rangle_M = \int_{-\infty}^{+\infty} x\left\{i\left[1 - |\psi(x)|^2\right]\right\} dx = i\int_{x_c-\frac{L}{2}}^{x_c+\frac{L}{2}} x\left\{1 - \frac{2}{L}\sin^2\left[k_n\left(x - x_c + \frac{L}{2}\right)\right]\right\} dx$$

$$= i\left\{\int_{x_c-\frac{L}{2}}^{x_c+\frac{L}{2}} x dx - \int_{x_c-\frac{L}{2}}^{x_c+\frac{L}{2}} \frac{2}{L} x \sin^2\left[k_n\left(x - x_c + \frac{L}{2}\right)\right] dx\right\} = i\left\{\left[\frac{x^2}{2}\right]_{x_c-\frac{L}{2}}^{x_c+\frac{L}{2}} - \langle x\rangle_R\right\}$$

$$= i\left\{\left[\frac{\left(x_c + \frac{L}{2}\right)^2}{2} - \frac{\left(x_c - \frac{L}{2}\right)^2}{2}\right] - x_c\right\}$$

$$= i(x_c L - x_c) = i x_c(L - 1)$$

To simplify, consider here and in what follows that $x_c = 0 \Leftrightarrow \langle x\rangle_R = 0$ and $\langle x\rangle_M = 0$.

Moreover,

$$\mathrm{Var}_{x,M} = \langle x^2 \rangle_M - \langle x \rangle_M^2 = \left\{ \int_{-\infty}^{+\infty} x^2 \left\{ i\left[1 - |\psi(x)|^2\right] \right\} dx \right\} - \left\{ \int_{-\infty}^{+\infty} x \left\{ i\left[1 - |\psi(x)|^2\right] \right\} dx \right\}^2$$

$$= i \int_{x_c - \frac{L}{2}}^{x_c + \frac{L}{2}} x^2 \left\{ 1 - \frac{2}{L} \sin^2 \left[k_n \left(x - x_c + \frac{L}{2} \right) \right] \right\} dx - 0$$

$$= i \left\{ \int_{x_c - \frac{L}{2}}^{x_c + \frac{L}{2}} x^2 dx - \int_{x_c - \frac{L}{2}}^{x_c + \frac{L}{2}} x^2 \left\{ \frac{2}{L} \sin^2 \left[k_n \left(x - x_c + \frac{L}{2} \right) \right] \right\} dx \right\}$$

$$= i \left\{ \int_{-\frac{L}{2}}^{+\frac{L}{2}} u^2 du - \mathrm{Var}_{x,R} \right\} = i \left\{ \left[\frac{u^3}{3} \right]_{-\frac{L}{2}}^{+\frac{L}{2}} - \mathrm{Var}_{x,R} \right\} = i \left\{ \frac{L^3}{12} - \frac{L^2}{12} \left(1 - \frac{6}{n^2 \pi^2} \right) \right\}$$

$$= i \left\{ \frac{L^2}{12} \left[L - \left(1 - \frac{6}{n^2 \pi^2} \right) \right] \right\}$$

In the probability set and the universe $\mathcal{C} = \mathcal{R} + \mathcal{M}$, we have from *CPP*:

$$\langle x \rangle_C = \int_{-\infty}^{+\infty} x[z(x)] dx = \int_{-\infty}^{+\infty} x \left\{ |\psi(x)|^2 + i\left[1 - |\psi(x)|^2\right] \right\} dx$$

$$= \int_{-\infty}^{+\infty} x |\psi(x)|^2 dx + \int_{-\infty}^{+\infty} xi\left[1 - |\psi(x)|^2\right] dx$$

$$= \int_{x_c - \frac{L}{2}}^{x_c + \frac{L}{2}} x \frac{2}{L} \sin^2 \left[k_n \left(x - x_c + \frac{L}{2} \right) \right] dx + i \int_{x_c - \frac{L}{2}}^{x_c + \frac{L}{2}} x \left\{ 1 - \frac{2}{L} \sin^2 \left[k_n \left(x - x_c + \frac{L}{2} \right) \right] \right\} dx$$

$$= \langle x \rangle_R + \langle x \rangle_M = x_c + ix_c(L-1) = x_c[1 + i(L-1)] = 0 \text{ for } x_c = 0$$

$$\mathrm{Var}_{x,C} = \langle x^2 \rangle_C - \langle x \rangle_C^2 = \left[\int_{-\infty}^{+\infty} x^2 [z(x)] dx \right] - \left[\langle x \rangle_R + \langle x \rangle_M \right]^2$$

$$= \left[\int_{-\infty}^{+\infty} x^2 \left\{ |\psi(x)|^2 + i\left[1 - |\psi(x)|^2\right] \right\} dx \right] - \left[\langle x \rangle_R + \langle x \rangle_M \right]^2$$

$$= \left[\int_{-\infty}^{+\infty} x^2 |\psi(x)|^2 dx + \int_{-\infty}^{+\infty} x^2 i\left[1 - |\psi(x)|^2\right] dx \right] - \left[\langle x \rangle_R + \langle x \rangle_M \right]^2$$

$$= \left[\langle x^2\rangle_R + \langle x^2\rangle_M\right] - \left[\langle x\rangle_R + \langle x\rangle_M\right]^2 = \left[\langle x^2\rangle_R + \langle x^2\rangle_M\right] - \left[\langle x\rangle_R^2 + \langle x\rangle_M^2 + 2\langle x\rangle_R\langle x\rangle_M\right]$$
$$= \left[\langle x^2\rangle_R - \langle x\rangle_R^2\right] + \left[\langle x^2\rangle_M - \langle x\rangle_M^2\right] - 2\langle x\rangle_R\langle x\rangle_M = \mathrm{Var}_{x,R} + \mathrm{Var}_{x,M} - 2\langle x\rangle_R\langle x\rangle_M$$
$$= \frac{L^2}{12}\left(1 - \frac{6}{n^2\pi^2}\right) + i\left\{\frac{L^2}{12}\left[L - \left(1 - \frac{6}{n^2\pi^2}\right)\right]\right\} - 2(0)(0)$$
$$= \frac{L^2}{12}\left(1 - \frac{6}{n^2\pi^2}\right) + i\left\{\frac{L^2}{12}\left[L - \left(1 - \frac{6}{n^2\pi^2}\right)\right]\right\}$$

The following tables (**Tables 1–4**) compute the position distribution characteristics for $x_c = 0, L = 200$, and $n = 1,2,3,20$.

Position distribution characteristics	$x_c = 0, L = 200, n = 1$
$\langle x\rangle_R$	0
$\mathrm{Var}_{x,R}$	1.3069e+03
$\langle x\rangle_M$	0
$\mathrm{Var}_{x,M}$	i×6.6536e+05
$\langle x\rangle_C = \langle x\rangle_R + \langle x\rangle_M$	0+i(0)
$\mathrm{Var}_{x,C} = \mathrm{Var}_{x,R} + \mathrm{Var}_{x,M} - 2\langle x\rangle_R\langle x\rangle_M$	1.3069e+03+i×6.6536e+05

Table 1.
The position distribution characteristics for $x_c = 0$, $L = 200$, and $n = 1$.

Position distribution characteristics	$x_c = 0, L = 200, n = 2$
$\langle x\rangle_R$	0
$\mathrm{Var}_{x,R}$	2.8267e+03
$\langle x\rangle_M$	0
$\mathrm{Var}_{x,M}$	i×6.6384e+05
$\langle x\rangle_C = \langle x\rangle_R + \langle x\rangle_M$	0+i(0)
$\mathrm{Var}_{x,C} = \mathrm{Var}_{x,R} + \mathrm{Var}_{x,M} - 2\langle x\rangle_R\langle x\rangle_M$	2.8267e+03+i×6.6384e+05

Table 2.
The position distribution characteristics for $x_c = 0$, $L = 200$, and $n = 2$.

Position distribution characteristics	$x_c = 0, L = 200, n = 3$
$\langle x\rangle_R$	0
$\mathrm{Var}_{x,R}$	3.1082e+03
$\langle x\rangle_M$	0
$\mathrm{Var}_{x,M}$	i×6.6356e+05
$\langle x\rangle_C = \langle x\rangle_R + \langle x\rangle_M$	0+i(0)
$\mathrm{Var}_{x,C} = \mathrm{Var}_{x,R} + \mathrm{Var}_{x,M} - 2\langle x\rangle_R\langle x\rangle_M$	3.1082e+03+i×6.6356e+05

Table 3.
The position distribution characteristics for $x_c = 0$, $L = 200$, and $n = 3$.

Position distribution characteristics	$x_c = 0, L = 200, n = 20$
$\langle x \rangle_R$	0
$\text{Var}_{x,R}$	3.3283e+03
$\langle x \rangle_M$	0
$\text{Var}_{x,M}$	$i\times$6.6334e+05
$\langle x \rangle_C = \langle x \rangle_R + \langle x \rangle_M$	0+i(0)
$\text{Var}_{x,C} = \text{Var}_{x,R} + \text{Var}_{x,M} - 2\langle x \rangle_R \langle x \rangle_M$	3.3283e+03+$i\times$6.6334e+05

Table 4.
The position distribution characteristics for $x_c = 0$, $L = 200$, and $n = 20$.

For $n \gg 1$ (large n) and with $x_c = 0$ we get:

$$\text{Var}_{x,R} \to \frac{L^2}{12} = 3.3333\ldots\text{e} + 03,$$

$$\text{Var}_{x,M} \to i\left\{\frac{L^2(L-1)}{12}\right\} = i \times 6.6333\ldots\text{e} + 05$$

$$\text{Var}_{x,C} \to \frac{L^2}{12} + i\left\{\frac{L^2(L-1)}{12}\right\} - 2(0)(0) = 3.3333\ldots\text{e} + 03 + i \times 6.6333\ldots\text{e} + 05$$

6. Conclusion and perspectives

In the current research work, the original extended model of eight axioms (*EKA*) of A. N. Kolmogorov was connected and applied to the infinite potential well problem in quantum mechanics theory. Thus, a tight link between quantum mechanics and the novel paradigm (*CPP*) was achieved. Consequently, the model of "Complex Probability" was more developed beyond the scope of my 19 previous research works on this topic.

Additionally, as it was proved and verified in the novel model, before the beginning of the random phenomenon simulation and at its end we have the chaotic factor (*Chf* and *MChf*) is zero, and the degree of our knowledge (*DOK*) is one since the stochastic fluctuations and effects have either not started yet or they have terminated and finished their task on the probabilistic phenomenon. During the execution of the nondeterministic phenomenon and experiment, we also have: $0.5 \leq DOK < 1$, $-0.5 \leq Chf < 0$, and $0 < MChf \leq 0.5$. We can see that during this entire process we have incessantly and continually $Pc^2 = DOK - Chf = DOK + MChf = 1 = Pc$, which means that the simulation which behaved randomly and stochastically in the real set and universe $\mathcal{R}$ is now certain and deterministic in the complex probability set and universe $\mathcal{C} = \mathcal{R} + \mathcal{M}$, and this after adding to the random experiment executed in the real universe $\mathcal{R}$, the contributions of the imaginary set and universe $\mathcal{M}$, and hence, after eliminating and subtracting the chaotic factor from the degree of our knowledge. Furthermore, the real, imaginary, complex, and deterministic probabilities that correspond to each value of the position random variable X have been determined in the three probabilities sets and universes, which are $\mathcal{R}$, $\mathcal{M}$, and $\mathcal{C}$ by P_r, P_m, Z and Pc, respectively. Consequently, at each value of X, the novel quantum mechanics and *CPP* parameters P_r, P_m, P_m/i, *DOK*, *Chf*, *MChf*, *Pc*, and Z are surely and perfectly predicted in the complex probabilities set and universe $\mathcal{C}$ with *Pc* maintained equal to one permanently and repeatedly.

In addition, referring to all these obtained graphs and executed simulations throughout the whole research work, we are able to quantify and visualize both the system chaos and stochastic effects and influences (expressed and materialized by *Chf* and *MChf*) and the certain knowledge (expressed and materialized by *DOK* and *Pc*) of the new paradigm. This is without any doubt very fruitful, wonderful, and fascinating and proves and reveals once again the advantages of extending A. N. Kolmogorov's five axioms of probability, and hence, the novelty and benefits of my inventive and original model in the fields of prognostics, applied mathematics, and quantum mechanics that can be called verily: "The Complex Probability Paradigm."

As a future and prospective research and challenges, we aim to develop the novel prognostic paradigm conceived and implement it in a large set of random and nondeterministic phenomena in quantum mechanics theory.

Nomenclature

$\mathcal{R}$	real set of events and probabilities.
$\mathcal{M}$	imaginary set of events and probabilities.
$\mathcal{C}$	complex set of events and probabilities.
i	the imaginary number where $i = \sqrt{-1}$ or $i^2 = -1$
EKA	Extended Kolmogorov's Axioms.
CPP	complex probability paradigm.
P_{rob}	probability of any event.
P_r	probability in the real set $\mathcal{R}$
P_m	probability in the imaginary set $\mathcal{M}$ corresponding to the real probability in $\mathcal{R}$.
Pc	probability of an event in $\mathcal{R}$ with its associated complementary event in $\mathcal{M}$ = probability in the complex probability set $\mathcal{C}$.
Z	complex probability number = sum of P_r and P_m = complex random vector
$DOK = \lvert Z\rvert^2$	the degree of our knowledge of the random system or experiment, it is the square of the norm of Z.
Chf	the chaotic factor of Z
MChf	magnitude of the chaotic factor of Z
$\lvert\psi(x)\rvert^2$	wave function position probability density function.
$\lvert\phi(p)\rvert^2$	wave function momentum probability density function.
$\langle x\rangle_R, \langle x\rangle_M, \langle x\rangle_C$	means, expectations, or averages of the wave function position probability distribution function in $\mathcal{R}$, $\mathcal{M}$, and $\mathcal{C}$, respectively.
$\mathrm{Var}_{x,R}, \mathrm{Var}_{x,M}, \mathrm{Var}_{x,C}$	variances of the wave function position probability distribution function in $\mathcal{R}$, $\mathcal{M}$, and $\mathcal{C}$, respectively.
$\langle p\rangle_R, \langle p\rangle_M, \langle p\rangle_C$	means, expectations, or averages of the wave function momentum probability distribution function in $\mathcal{R}$, $\mathcal{M}$, and $\mathcal{C}$, respectively.
$\mathrm{Var}_{p,R}, \mathrm{Var}_{p,M}, \mathrm{Var}_{p,C}$	variances of the wave function momentum probability distribution function in $\mathcal{R}$, $\mathcal{M}$, and $\mathcal{C}$, respectively.

H_x^R	particle position entropy in the real universe $\mathcal{R}$.
$NegH_x^R$	particle position negative entropy in the real universe $\mathcal{R}$.
$\overline{H}_x^R$	particle position complementary entropy in the real universe $\mathcal{R}$.
H_x^M	particle position entropy in the imaginary universe $\mathcal{M}$.
H_x^C	particle position entropy in the complex universe $\mathcal{C}$.
H_p^R	particle momentum entropy in the real universe $\mathcal{R}$.
$NegH_p^R$	particle momentum negative entropy in the real universe $\mathcal{R}$.
$\overline{H}_p^R$	particle momentum complementary entropy in the real universe $\mathcal{R}$.
H_p^M	particle momentum entropy in the imaginary universe $\mathcal{M}$.
H_p^C	particle momentum entropy in the complex universe $\mathcal{C}$.

Author details

Abdo Abou Jaoudé
Faculty of Natural and Applied Sciences, Department of Mathematics and Statistics, Notre Dame University-Louaize, Lebanon

*Address all correspondence to: abdoaj@idm.net.lb

References

[1] Wikipedia, the free encyclopedia, Quantum Mechanics. https://en.wikipedia.org/

[2] Wikipedia, the free encyclopedia, Particle in a Box. https://en.wikipedia.org/

[3] Cartwright N. How the Laws of Physics Lie. Oxford, United Kingdom: Oxford University Press; 1983. p. 142. https://philpapers.org/rec/CARHTL

[4] Abou Jaoude A, El-Tawil K, Kadry S. Prediction in complex dimension using Kolmogorov's set of axioms. Journal of Mathematics and Statistics, Science Publications. 2010;**6**(2):116-124

[5] Abou Jaoude A. The complex statistics paradigm and the law of large numbers. Journal of Mathematics and Statistics, Science Publications. 2013;**9**(4):289-304

[6] Abou Jaoude A. The theory of complex probability and the first order reliability method. Journal of Mathematics and Statistics, Science Publications. 2013;**9**(4):310-324

[7] Abou Jaoude A. Complex probability theory and prognostic. Journal of Mathematics and Statistics, Science Publications. 2014;**10**(1):1-24

[8] Abou Jaoude A. The complex probability paradigm and analytic linear prognostic for vehicle suspension systems. American Journal of Engineering and Applied Sciences, Science Publications. 2015;**8**(1):147-175

[9] Abou Jaoude A. The paradigm of complex probability and the Brownian motion. Systems Science and Control Engineering. 2015;**3**(1):478-503

[10] Abou Jaoude A. The paradigm of complex probability and Chebyshev's inequality. Systems Science and Control Engineering. 2016;**4**(1):99-137

[11] Abou Jaoude A. The paradigm of complex probability and analytic nonlinear prognostic for vehicle suspension systems. Systems Science and Control Engineering. 2016;**4**(1):99-137

[12] Abou Jaoude A. The paradigm of complex probability and analytic linear prognostic for unburied petrochemical pipelines. Systems Science and Control Engineering, Taylor and Francis Publishers. 2017;**5**(1):178-214

[13] Abou Jaoude A. The paradigm of complex probability and Claude Shannon's information theory. Systems Science and Control Engineering. 2017;**5**(1):380-425

[14] Abou Jaoude A. The paradigm of complex probability and analytic nonlinear prognostic for unburied petrochemical pipelines. Systems Science and Control Engineering. 2017;**5**(1): 495-534

[15] Abou Jaoude A. The paradigm of complex probability and Ludwig Boltzmann's Entropy. Systems Science and Control Engineering. 2018;**6**(1):108-149

[16] Abou Jaoude A. The paradigm of complex probability and Monte Carlo methods. Systems Science and Control Engineering. 2019;**7**(1):407-451

[17] Abou Jaoude A. Analytic prognostic in the linear damage case applied to buried petrochemical pipelines and the complex probability paradigm. Fault Detection, Diagnosis and Prognosis. 2020;**1**(5):65-103. DOI: 10.5772/intechopen.90157

[18] Abou Jaoude A. The Monte Carlo techniques and the complex probability

paradigm. In: Forecasting in Mathematics - Recent Advances, New Perspectives and Applications. London, UK: IntechOpen; 2020

[19] Abou Jaoude A. The paradigm of complex probability and prognostic using FORM. London Journal of Research in Science: Natural and Formal (LJRS). 2020;**20**(4):1-65

[20] Abou Jaoude A. The paradigm of complex probability and the central limit theorem. London Journal of Research in Science: Natural and Formal (LJRS). 2020;**20**(5):1-57

[21] Abou Jaoude A. The paradigm of complex probability and Thomas Bayes' Theorem. In: The Monte Carlo Methods - Recent Advances, New Perspectives and Applications. London, UK: IntechOpen; 2021

[22] Abou Jaoude A. The paradigm of complex probability and Isaac Newton's Classical mechanics: On the foundation of statistical physics. In: The Monte Carlo Methods - Recent Advances, New Perspectives and Applications. London, UK: IntechOpen; 2021

[23] Benton W. Probability, Encyclopedia Britannica. Chicago: Encyclopedia Britannica Inc; 1966

[24] Benton W. Mathematical Probability, Encyclopedia Britannica. Chicago: Encyclopedia Britannica Inc; 1966

[25] Feller W. An Introduction to Probability Theory and Its Applications. 3rd ed. New York: Wiley; 1968

[26] Walpole R, Myers R, Myers S, Ye K. Probability and Statistics for Engineers and Scientists. 7th ed. New Jersey: Prentice Hall; 2002

[27] Freund JE. Introduction to Probability. New York: Dover Publications; 1973

[28] Srinivasan SK, Mehata KM. Stochastic Processes. 2nd ed. New Delhi: McGraw-Hill; 1988

[29] Aczel AD. God's Equation. New York, New York, United States: Dell Publishing; 2000

[30] Barrow JD. The Book of Nothing. New York, New York, United States: Vintage; 2002

[31] Barrow JD. New Theories of Everything. Oxford, United Kingdom: Oxford University Press; 2007

[32] Barrow JD. The Infinite Book. New York, New York, United States: Vintage; 2006

[33] Becker K, Becker M, Schwarz JH. String Theory and M-Theory. Cambridge, United Kingdom: Cambridge University Press; 2007

[34] De Broglie L. La Physique Nouvelle et les Quanta. Paris, France: Flammarion; 1937

[35] Einstein A. Traduction Française: Comment Je Vois le Monde. Paris, France: Flammarion; 1979

[36] Einstein A. La Relativité. Paris, France: Petite Bibliothèque Payot; 2001

[37] Feynmann R. Traduction Française: La Nature de la Physique. Paris: Le Seuil; 1980

[38] Balibar F. Albert Einstein: Physique, Philosophie, Politique. Paris: Le Seuil; 2002

[39] Gates E. Einstein's Telescope. Manhattan, United States: Norton; 2010

[40] Greene B. The Elegant Universe. New York, New York, United States: Vintage; 2003

[41] Greene B. The Fabric of the Cosmos. New York, New York, United States: Vintage; 2004

[42] Gribbin J. Traduction Française: A la Poursuite du Big Bang. Paris, France: Flammarion; 1994

[43] Gubser SS. The Little Book of String Theory. Princeton, New Jersey, United States: Princeton; 2010

[44] Hawking S. Traduction Française: Trous Noirs et Bébés Univers. Paris: René Lambert; 2000

[45] Hawking S. Traduction Française: Une Brève Histoire du Temps: Du Big Bang aux Trous Noirs. Isabelle Naddeo-Souriau: Flammarion; 1989

[46] Hawking S. The Essential Einstein, His Greatest Works. New York, New York, United States: Penguin Books; 2007

[47] Heath TL. The Elements of Euclid. New York, United States: Dover; 1956

[48] Hoffmann B, Dukas H. Albert Einstein, Creator and Rebel. New York: Viking; 1972

[49] Luminet J-P. Les Trous Noirs. Paris, France: Le Seuil; 1992

[50] Nicolson I. Dark Side of the Universe. United States: Johns Hopkins; 2007

[51] Panek R. The 4% Universe. United States: First Mariner Books; 2011

[52] Penrose R. Traduction Française: Les Deux Infinis et L'Esprit Humain. Paris, France: Flammarion; 1999

[53] Penrose R. Cycles of Time. New York, United States: Vintage; 2011

[54] Penrose R. The Road to Reality. New York, United States: Vintage; 2004

[55] Planck M. Traduction Française: Initiations à la Physique. Paris, France: Flammarion; 1993

[56] Poincaré H. La Science et L'Hypothèse. Paris: Flammarion; 1968

[57] Proust D, Vanderriest C. Les Galaxies et la Structure de L'Univers. Paris, France: Le Seuil; 1997

[58] Reeves H. Patience dans L'Azur. Paris, France: Le Seuil; 1988

[59] Ronan C. Traduction Française: Histoire Mondiale des Sciences. Paris, France: Le Seuil; 1988

[60] Sagan C. Traduction Française: Cosmic Connection ou L'appel des étoiles. Paris, France: Vincent Bardet; 1975

[61] Singh S. Big Bang – The Origin of The Universe. United States: Harper Perennial; 2005

[62] Thorne KS. Traduction Française: Trous Noirs et Distorsions du Temps. Paris, France: Flammarion; 1997

[63] Weinberg S. Traduction Française: Les Trois Premières Minutes de L'Univers. Paris, France: Jean-Benoit Yelnik; 1988

[64] Weinberg S. Cosmology. Oxford, United Kingdom: Oxford University Press; 2008

[65] Weinberg S. Dreams of a Final Theory. New York, United States: Vintage; 1993

[66] Stewart I. Does God Play Dice? 2nd ed. Oxford: Blackwell Publishing; 2002

[67] Stewart I. From Here to Infinity. 2nd ed. Oxford: Oxford University Press; 1996

[68] Stewart I. In Pursuit of the Unknown. New York: Basic Books; 2012

[69] Barrow J. Pi in the Sky. Oxford: Oxford University Press; 1992

[70] Bogdanov I, Bogdanov G. Au Commencement du Temps. Paris: Flammarion; 2009

[71] Bogdanov I, Bogdanov G. Le Visage de Dieu. Paris: Editions Grasset et Fasquelle; 2010

[72] Bogdanov I, Bogdanov G. La Pensée de Dieu. Paris: Editions Grasset et Fasquelle; 2012

[73] Bogdanov I, Bogdanov G. La Fin du Hasard. Paris: Editions Grasset et Fasquelle; 2013

[74] Van Kampen NG. Stochastic Processes in Physics and Chemistry. Sydney: Elsevier; 2006

[75] Bell ET. The Development of Mathematics. New York: Dover Publications, Inc.; 1992

[76] Boursin J-L. Les Structures du Hasard. Paris: Editions du Seuil; 1986

[77] Dacunha-Castelle D. Chemins de l'Aléatoire. Paris: Flammarion; 1996

[78] Dalmedico-Dahan A, Chabert J-L, Chemla K. Chaos Et Déterminisme. Paris: Edition du Seuil; 1992

[79] Ekeland I. Au Hasard. In: La Chance, la Science et le Monde. Paris: Editions du Seuil; 1991

[80] Gleick J. Chaos, Making a New Science. New York: Penguin Books; 1997

[81] Dalmedico-Dahan A, Peiffer J. Une Histoire des Mathématiques. Paris: Edition du Seuil; 1986

[82] Gullberg J. Mathematics from the Birth of Numbers. New York: W.W. Norton & Company; 1997

[83] Science Et Vie. Le Mystère des Mathématiques. Numéro 984. 1999

[84] Davies P. The Mind of God. London: Penguin Books; 1993

[85] Gillies D. Philosophical Theories of Probability. London: Routledge; 2000

[86] Guillen M. Initiation Aux Mathématiques. Paris: Albin Michel; 1995

[87] Hawking S. On the Shoulders of Giants. London: Running Press; 2002

[88] Hawking S. God Created the Integers. London: Penguin Books; 2005

[89] Hawking S. The Dreams that Stuff is Made of. London: Running Press; 2011

[90] Pickover C. Archimedes to Hawking. Oxford: Oxford University Press; 2008

[91] Abou Jaoude A. The Computer Simulation of Monté Carlo Methods and Random Phenomena. United Kingdom: Cambridge Scholars Publishing; 2019

[92] Abou Jaoude A. The Analysis of Selected Algorithms for the Stochastic Paradigm. United Kingdom: Cambridge Scholars Publishing; 2019

[93] Abou Jaoude A. The Analysis of Selected Algorithms for the Statistical Paradigm. Vol. 1. The Republic of Moldova: Generis Publishing; 2021

[94] Abou Jaoude A. The Analysis of Selected Algorithms for the Statistical Paradigm. Vol. 2. The Republic of Moldova: Generis Publishing; 2021

[95] Abou Jaoude A. Forecasting in Mathematics – Recent Advances, New Perspectives and Applications. London: IntechOpen; 2021

[96] Abou Jaoude A. The Monte Carlo Methods - Recent Advances, New Perspectives and Applications. London: IntechOpen; 2022

[97] Abou Jaoude A. Applied Mathematics: Numerical Methods and Algorithms for Applied Mathematicians. Bircham: Bircham International University; 2004

[98] Abou Jaoude A. Computer Simulation of Monté Carlo Methods and Random Phenomena. Bircham: Bircham International University; 2005

[99] Abou Jaoude A. Analysis and Algorithms for the Statistical and Stochastic Paradigm. Bircham: Bircham International University; 2007

Chapter 2

The Paradigm of Complex Probability and Quantum Mechanics: The Infinite Potential Well Problem – The Momentum Wavefunction and the Wavefunction Entropies

Abdo Abou Jaoudé

Abstract

The mathematical probability concept was set forth by Andrey Nikolaevich Kolmogorov in 1933 by laying down a five-axioms system. This scheme can be improved to embody the set of imaginary numbers after adding three new axioms. Accordingly, any stochastic phenomenon can be performed in the set $\mathcal{C}$ of complex probabilities which is the summation of the set $\mathcal{R}$ of real probabilities and the set $\mathcal{M}$ of imaginary probabilities. Our objective now is to encompass complementary imaginary dimensions to the stochastic phenomenon taking place in the "real" laboratory in $\mathcal{R}$ and as a consequence to calculate in the sets $\mathcal{R}$, $\mathcal{M}$, and $\mathcal{C}$ all the corresponding probabilities. Hence, the probability is permanently equal to one in the entire set $\mathcal{C} = \mathcal{R} + \mathcal{M}$ independently of all the probabilities of the input stochastic variable distribution in $\mathcal{R}$, and subsequently, the output of the random phenomenon in $\mathcal{R}$ can be determined perfectly in $\mathcal{C}$. This is due to the fact that the probability in $\mathcal{C}$ is calculated after the elimination and subtraction of the chaotic factor from the degree of our knowledge of the nondeterministic phenomenon. My innovative Complex Probability Paradigm (*CPP*) will be applied to the established theory of quantum mechanics in order to express it completely deterministically in the universe $\mathcal{C} = \mathcal{R} + \mathcal{M}$.

Keywords: degree of our knowledge, chaotic factor, complex random vector, probability norm, complex probability set $\mathcal{C}$, momentum wavefunction, imaginary entropy, complex entropy

1. Introduction

1.1 The momentum wavefunction and *CPP*

1.1.1 The momentum wavefunction probability distribution and CPP

The probability density for finding a particle with a given momentum is derived from the wavefunction as $f(p) = |\phi(p)|^2$. As with position, the wavefunction momentum probability density function (*PDF*) for finding the particle at a given momentum depends upon its state, and is given by [1, 2]:

$$f(p) = |\phi(p)|^2 = \frac{L}{\pi\hbar}\left(\frac{n\pi}{n\pi + pL/\hbar}\right)^2 \operatorname{sinc}^2\left[\frac{1}{2}(n\pi - pL/\hbar)\right]$$

Where $\hbar = \frac{h}{2\pi}$ is the reduced Planck constant and $\operatorname{sinc}(x) = \frac{\sin(x)}{x}$ is the cardinal sine *sinc* function.

Therefore, the wavefunction momentum cumulative probability distribution function (*CDF*) which is equal to $P_r(P)$ in $\mathcal{R}$ is:

$$P_r(P) = F(p_j) = P_{rob}(P \le p_j) = \int_{-\infty}^{p_j} |\phi(p)|^2 dp$$

$$= \int_{-\infty}^{p_j} \frac{L}{\pi\hbar}\left(\frac{n\pi}{n\pi + pL/\hbar}\right)^2 \operatorname{sinc}^2\left[\frac{1}{2}(n\pi - pL/\hbar)\right] dp$$

And the real complementary probability to $P_r(P)$ in $\mathcal{R}$ which is $P_m(P)/i$ is:

$$P_m(P)/i = 1 - P_r(P) = 1 - F(p_j) = 1 - P_{rob}(P \le p_j) = P_{rob}(P > p_j)$$

$$= 1 - \int_{-\infty}^{p_j} |\phi(p)|^2 dp = \int_{p_j}^{+\infty} |\phi(p)|^2 dp$$

$$= 1 - \int_{-\infty}^{p_j} \frac{L}{\pi\hbar}\left(\frac{n\pi}{n\pi + pL/\hbar}\right)^2 \operatorname{sinc}^2\left[\frac{1}{2}(n\pi - pL/\hbar)\right] dp$$

$$= \int_{p_j}^{+\infty} \frac{L}{\pi\hbar}\left(\frac{n\pi}{n\pi + pL/\hbar}\right)^2 \operatorname{sinc}^2\left[\frac{1}{2}(n\pi - pL/\hbar)\right] dp$$

Consequently, the imaginary complementary probability to $P_r(P)$ in $\mathcal{M}$ which is $P_m(P)$ is:

$$P_m(P) = i[1 - P_r(P)] = i\left[1 - F\left(p_j\right)\right] = i\left[1 - P_{rob}\left(P \le p_j\right)\right] = iP_{rob}\left(P > p_j\right)$$

$$= i\left[1 - \int_{-\infty}^{p_j} |\phi(p)|^2 dp\right] = i\int_{p_j}^{+\infty} |\phi(p)|^2 dp$$

$$= i\left[1 - \int_{-\infty}^{p_j} \frac{L}{\pi\hbar}\left(\frac{n\pi}{n\pi + pL/\hbar}\right)^2 \operatorname{sinc}^2\left[\frac{1}{2}(n\pi - pL/\hbar)\right] dp\right]$$

$$= i\int_{p_j}^{+\infty} \frac{L}{\pi\hbar}\left(\frac{n\pi}{n\pi + pL/\hbar}\right)^2 \operatorname{sinc}^2\left[\frac{1}{2}(n\pi - pL/\hbar)\right] dp$$

Furthermore, the complex random number or vector in $\mathcal{C} = \mathcal{R} + \mathcal{M}$ which is $Z(P)$ is:

$$Z(P) = P_r(P) + P_m(P) = P_r(P) + i[1 - P_r(P)] = F\left(p_j\right) + i\left[1 - F\left(p_j\right)\right]$$

$$= P_{rob}\left(P \le p_j\right) + i\left[1 - P_{rob}\left(P \le p_j\right)\right] = P_{rob}\left(P \le p_j\right) + iP_{rob}\left(P > p_j\right)$$

$$= \int_{-\infty}^{p_j} |\phi(p)|^2 dp + i\left[1 - \int_{-\infty}^{p_j} |\phi(p)|^2 dp\right] = \int_{-\infty}^{p_j} |\phi(p)|^2 dp + i\int_{p_j}^{+\infty} |\phi(p)|^2 dp$$

$$= \int_{-\infty}^{p_j} \frac{L}{\pi\hbar}\left(\frac{n\pi}{n\pi + pL/\hbar}\right)^2 \operatorname{sinc}^2\left[\frac{1}{2}(n\pi - pL/\hbar)\right] dp$$

$$+ i\left[1 - \int_{-\infty}^{p_j} \frac{L}{\pi\hbar}\left(\frac{n\pi}{n\pi + pL/\hbar}\right)^2 \operatorname{sinc}^2\left[\frac{1}{2}(n\pi - pL/\hbar)\right] dp\right]$$

$$= \int_{-\infty}^{p_j} \frac{L}{\pi\hbar}\left(\frac{n\pi}{n\pi + pL/\hbar}\right)^2 \operatorname{sinc}^2\left[\frac{1}{2}(n\pi - pL/\hbar)\right] dp$$

$$+ i\int_{p_j}^{+\infty} \frac{L}{\pi\hbar}\left(\frac{n\pi}{n\pi + pL/\hbar}\right)^2 \operatorname{sinc}^2\left[\frac{1}{2}(n\pi - pL/\hbar)\right] dp$$

Additionally, the degree of our knowledge which is $DOK(P)$ is:

$$
\begin{aligned}
DOK(P) &= [P_r(P)]^2 + [P_m(P)/i]^2 = [P_r(P)]^2 + [1 - P_r(P)]^2 = \left[F\left(p_j\right)\right]^2 + \left[1 - F\left(p_j\right)\right]^2 \\
&= \left[P_{rob}\left(P \le p_j\right)\right]^2 + \left[1 - P_{rob}\left(P \le p_j\right)\right]^2 = \left[P_{rob}\left(P \le p_j\right)\right]^2 + \left[P_{rob}\left(P > p_j\right)\right]^2 \\
&= \left[\int_{-\infty}^{p_j} |\phi(p)|^2 dp\right]^2 + \left[1 - \int_{-\infty}^{p_j} |\phi(p)|^2 dp\right]^2 = \left[\int_{-\infty}^{p_j} |\phi(p)|^2 dp\right]^2 + \left[\int_{p_j}^{+\infty} |\phi(p)|^2 dp\right]^2 \\
&= \left[\int_{-\infty}^{p_j} \frac{L}{\pi\hbar}\left(\frac{n\pi}{n\pi + pL/\hbar}\right)^2 \operatorname{sinc}^2\left[\frac{1}{2}(n\pi - pL/\hbar)\right] dp\right]^2 \\
&+ \left[1 - \int_{-\infty}^{p_j} \frac{L}{\pi\hbar}\left(\frac{n\pi}{n\pi + pL/\hbar}\right)^2 \operatorname{sinc}^2\left[\frac{1}{2}(n\pi - pL/\hbar)\right] dp\right]^2 \\
&= \left[\int_{-\infty}^{p_j} \frac{L}{\pi\hbar}\left(\frac{n\pi}{n\pi + pL/\hbar}\right)^2 \operatorname{sinc}^2\left[\frac{1}{2}(n\pi - pL/\hbar)\right] dp\right]^2 \\
&+ \left[\int_{p_j}^{+\infty} \frac{L}{\pi\hbar}\left(\frac{n\pi}{n\pi + pL/\hbar}\right)^2 \operatorname{sinc}^2\left[\frac{1}{2}(n\pi - pL/\hbar)\right] dp\right]^2
\end{aligned}
$$

Moreover, the chaotic factor which is $Chf(P)$ is:

$$
\begin{aligned}
Chf(P) &= 2iP_r(P)P_m(P) \\
&= 2iP_r(P) \times i[1 - P_r(P)] = -2P_r(P)[1 - P_r(P)] = -2F\left(p_j\right)\left[1 - F\left(p_j\right)\right] \\
&= -2P_{rob}\left(P \le p_j\right)\left[1 - P_{rob}\left(P \le p_j\right)\right] = -2P_{rob}\left(P \le p_j\right)P_{rob}\left(P > p_j\right) \\
&= -2\int_{-\infty}^{p_j} |\phi(p)|^2 dp \times \left[1 - \int_{-\infty}^{p_j} |\phi(p)|^2 dp\right] = -2\int_{-\infty}^{p_j} |\phi(p)|^2 dp \times \int_{p_j}^{+\infty} |\phi(p)|^2 dp \\
&= -2\int_{-\infty}^{p_j} \frac{L}{\pi\hbar}\left(\frac{n\pi}{n\pi + pL/\hbar}\right)^2 \operatorname{sinc}^2\left[\frac{1}{2}(n\pi - pL/\hbar)\right] dp \\
&\times \left[1 - \int_{-\infty}^{p_j} \frac{L}{\pi\hbar}\left(\frac{n\pi}{n\pi + pL/\hbar}\right)^2 \operatorname{sinc}^2\left[\frac{1}{2}(n\pi - pL/\hbar)\right] dp\right] \\
&= -2\int_{-\infty}^{p_j} \frac{L}{\pi\hbar}\left(\frac{n\pi}{n\pi + pL/\hbar}\right)^2 \operatorname{sinc}^2\left[\frac{1}{2}(n\pi - pL/\hbar)\right] dp \\
&\times \int_{p_j}^{+\infty} \frac{L}{\pi\hbar}\left(\frac{n\pi}{n\pi + pL/\hbar}\right)^2 \operatorname{sinc}^2\left[\frac{1}{2}(n\pi - pL/\hbar)\right] dp
\end{aligned}
$$

In addition, the magnitude of the chaotic factor which is $MChf(P)$ is:

$$\begin{aligned}
MChf(P) &= |Chf(P)| = -2iP_r(P)P_m(P) = -2iP_r(P) \times i[1 - P_r(P)] \\
&= 2P_r(P)[1 - P_r(P)] = 2F\left(p_j\right)\left[1 - F\left(p_j\right)\right] \\
&= 2P_{rob}\left(P \le p_j\right)\left[1 - P_{rob}\left(P \le p_j\right)\right] = 2P_{rob}\left(P \le p_j\right)P_{rob}\left(P > p_j\right) \\
&= 2\int_{-\infty}^{p_j} |\phi(p)|^2 dp \times \left[1 - \int_{-\infty}^{p_j} |\phi(p)|^2 dp\right] = 2\int_{-\infty}^{p_j} |\phi(p)|^2 dp \times \int_{p_j}^{+\infty} |\phi(p)|^2 dp \\
&= 2\int_{-\infty}^{p_j} \frac{L}{\pi\hbar}\left(\frac{n\pi}{n\pi + pL/\hbar}\right)^2 \text{sinc}^2\left[\frac{1}{2}(n\pi - pL/\hbar)\right] dp \\
&\quad \times \left[1 - \int_{-\infty}^{p_j} \frac{L}{\pi\hbar}\left(\frac{n\pi}{n\pi + pL/\hbar}\right)^2 \text{sinc}^2\left[\frac{1}{2}(n\pi - pL/\hbar)\right] dp\right] \\
&= 2\int_{-\infty}^{p_j} \frac{L}{\pi\hbar}\left(\frac{n\pi}{n\pi + pL/\hbar}\right)^2 \text{sinc}^2\left[\frac{1}{2}(n\pi - pL/\hbar)\right] dp \\
&\quad \times \int_{p_j}^{+\infty} \frac{L}{\pi\hbar}\left(\frac{n\pi}{n\pi + pL/\hbar}\right)^2 \text{sinc}^2\left[\frac{1}{2}(n\pi - pL/\hbar)\right] dp
\end{aligned}$$

Finally, the real probability in the complex probability universe $\mathcal{C} = \mathcal{R} + \mathcal{M}$ which is $Pc(P)$ is:

$$\begin{aligned}
Pc^2(P) &= \{[P_r(P)] + [P_m(P)/i]\}^2 = \{[P_r(P)] + [1 - P_r(P)]\}^2 \\
&= \left\{\left[F\left(p_j\right)\right] + \left[1 - F\left(p_j\right)\right]\right\}^2 = \left\{P_{rob}\left(P \le p_j\right) + \left[1 - P_{rob}\left(P \le p_j\right)\right]\right\}^2 \\
&= \left\{P_{rob}\left(P \le p_j\right) + P_{rob}\left(P > p_j\right)\right\}^2 \\
&= \left\{\int_{-\infty}^{p_j} |\phi(p)|^2 dp + \left[1 - \int_{-\infty}^{p_j} |\phi(p)|^2 dp\right]\right\}^2 \\
&= \left\{\int_{-\infty}^{p_j} |\phi(p)|^2 dp + \int_{p_j}^{+\infty} |\phi(p)|^2 dp\right\}^2 = \left\{\int_{-\infty}^{+\infty} |\phi(p)|^2 dp\right\}^2 \\
&= \left\{\begin{array}{l} \displaystyle\int_{-\infty}^{p_j} \frac{L}{\pi\hbar}\left(\frac{n\pi}{n\pi + pL/\hbar}\right)^2 \text{sinc}^2\left[\frac{1}{2}(n\pi - pL/\hbar)\right] dp \\ \displaystyle\quad + \left[1 - \int_{-\infty}^{p_j} \frac{L}{\pi\hbar}\left(\frac{n\pi}{n\pi + pL/\hbar}\right)^2 \text{sinc}^2\left[\frac{1}{2}(n\pi - pL/\hbar)\right] dp\right] \end{array}\right\}^2 \\
&= \left\{\begin{array}{l} \displaystyle\int_{-\infty}^{p_j} \frac{L}{\pi\hbar}\left(\frac{n\pi}{n\pi + pL/\hbar}\right)^2 \text{sinc}^2\left[\frac{1}{2}(n\pi - pL/\hbar)\right] dp \\ \displaystyle\quad + \int_{p_j}^{+\infty} \frac{L}{\pi\hbar}\left(\frac{n\pi}{n\pi + pL/\hbar}\right)^2 \text{sinc}^2\left[\frac{1}{2}(n\pi - pL/\hbar)\right] dp \end{array}\right\}^2 \\
&= \left\{\int_{-\infty}^{+\infty} \frac{L}{\pi\hbar}\left(\frac{n\pi}{n\pi + pL/\hbar}\right)^2 \text{sinc}^2\left[\frac{1}{2}(n\pi - pL/\hbar)\right] dp\right\}^2 = 1^2 = 1 = Pc(P)
\end{aligned}$$

And, $Pc(P)$ can be computed using CPP as follows:

$$
\begin{aligned}
Pc^2(P) &= DOK(P) - Chf(P) = [P_r(P)]^2 + [P_m(P)/i]^2 - 2iP_r(P)P_m(P) \\
&= [P_r(P)]^2 + [1 - P_r(P)]^2 + 2P_r(P)[1 - P_r(P)] = \{P_r(P) + [1 - P_r(P)]\}^2 \\
&= \left\{ \int_{-\infty}^{p_j} |\phi(p)|^2 dp + \left[1 - \int_{-\infty}^{p_j} |\phi(p)|^2 dp \right] \right\}^2 = \left\{ \int_{-\infty}^{p_j} |\phi(p)|^2 dp + \int_{p_j}^{+\infty} |\phi(p)|^2 dp \right\}^2 \\
&= \left\{ \int_{-\infty}^{+\infty} |\phi(p)|^2 dp \right\}^2 \\
&= 1^2 = 1 = Pc(P)
\end{aligned}
$$

And, $Pc(P)$ can be computed using always CPP as follows also:

$$
\begin{aligned}
Pc^2(P) &= DOK(P) + MChf(P) = [P_r(P)]^2 + [P_m(P)/i]^2 + [-2iP_r(P)P_m(P)] \\
&= [P_r(P)]^2 + [1 - P_r(P)]^2 + 2P_r(P)[1 - P_r(P)] = \{P_r(P) + [1 - P_r(P)]\}^2 \\
&= \left\{ \int_{-\infty}^{p_j} |\phi(p)|^2 dp + \left[1 - \int_{-\infty}^{p_j} |\phi(p)|^2 dp \right] \right\}^2 \\
&= \left\{ \int_{-\infty}^{p_j} |\phi(p)|^2 dp + \int_{p_j}^{+\infty} |\phi(p)|^2 dp \right\}^2 = \left\{ \int_{-\infty}^{+\infty} |\phi(p)|^2 dp \right\}^2 = 1^2 = 1 = Pc(P)
\end{aligned}
$$

Hence, the prediction of all the wavefunction momentum probabilities of the random infinite potential well problem in the universe $\mathcal{C} = \mathcal{R} + \mathcal{M}$ is permanently certain and perfectly deterministic.

1.1.2 The new model simulations

The following figures (**Figures 1–37**) illustrate all the calculations done above.

1.1.2.1 Simulations interpretation

In **Figures 1, 6, 11, 16, 21, 26, 31, 36, and 37** we can see the graphs of the probability density functions (PDF) of the wavefunction momentum probability distribution for this problem as functions of the random variable P for n = 1, 2, 3, 4, 5, 6, 7, 12, 100.

In **Figures 2, 7, 12, 17, 22, 27, and 32** we can see also the graphs and the simulations of all the CPP parameters (Chf, $MChf$, DOK, P_r, P_m/i, Pc) as functions of the random variable P for the wavefunction momentum probability distribution of the infinite potential well problem for n = 1, 2, 3, 4, 5, 6, 7. Hence, we can visualize all the new paradigm functions for this problem.

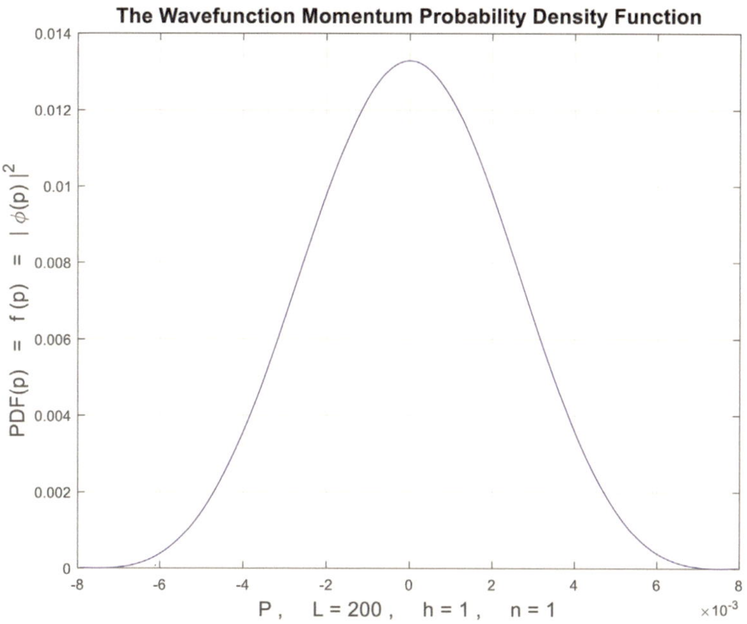

Figure 1.
The graph of the PDF *of the wavefunction momentum probability distribution as a function of the random variable* P *for* n = 1.

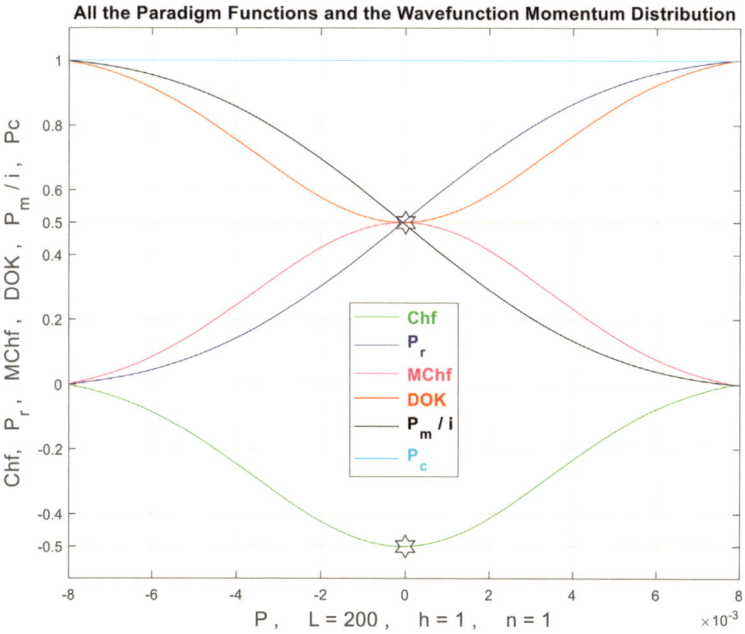

Figure 2.
The graphs of all the CPP *parameters as functions of the random variable* P *for the wavefunction momentum probability distribution for* n = 1.

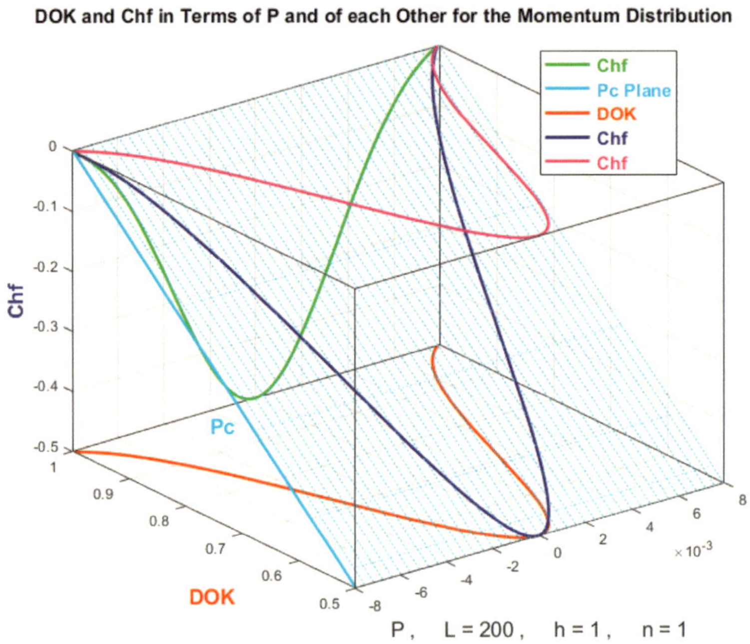

Figure 3.
The graphs of DOK *and* Chf *and the deterministic probability* Pc *in terms of* P *and of each other for the wavefunction momentum probability distribution for* n = 1.

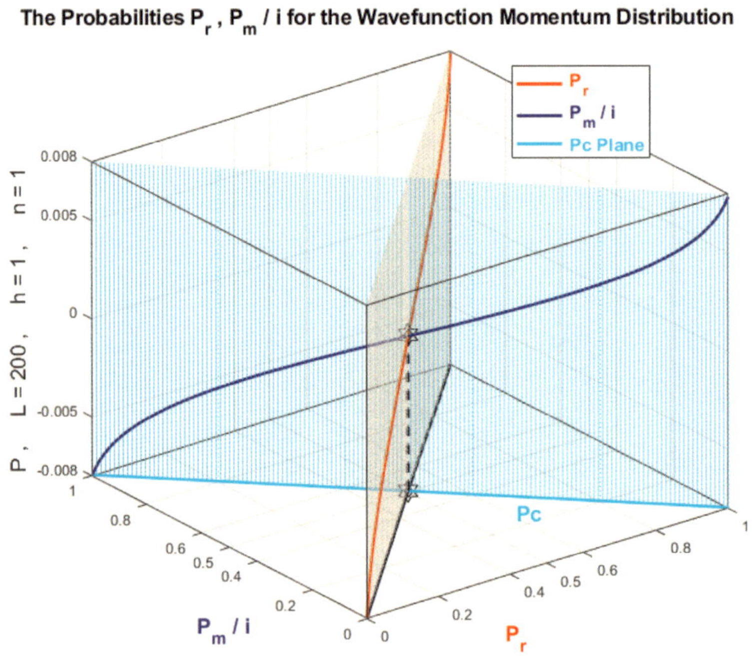

Figure 4.
The graphs of P_r *and* P_m/i *and* Pc *in terms of* P *and of each other for the wavefunction momentum probability distribution for* n = 1.

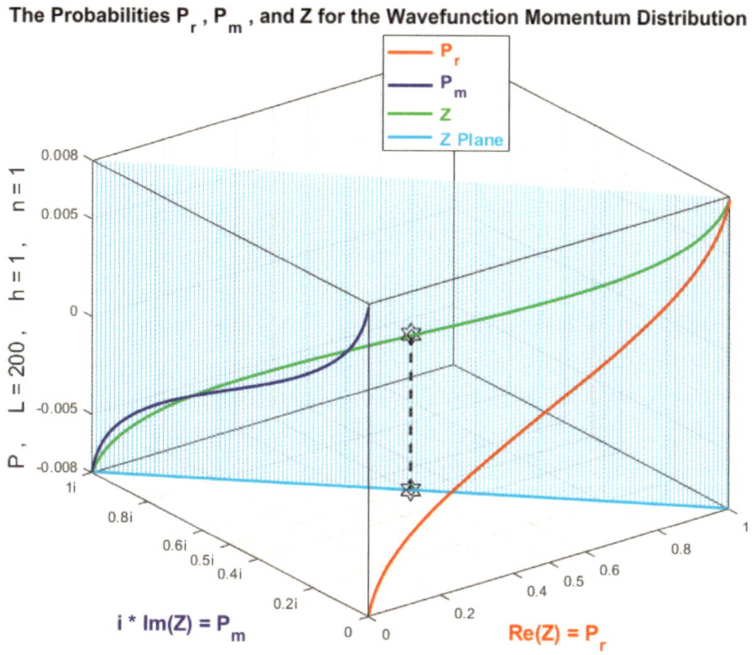

Figure 5.
The graphs of the probabilities P_r *and* P_m *and* Z *in terms of* P *for the wavefunction momentum probability distribution for* n = 1.

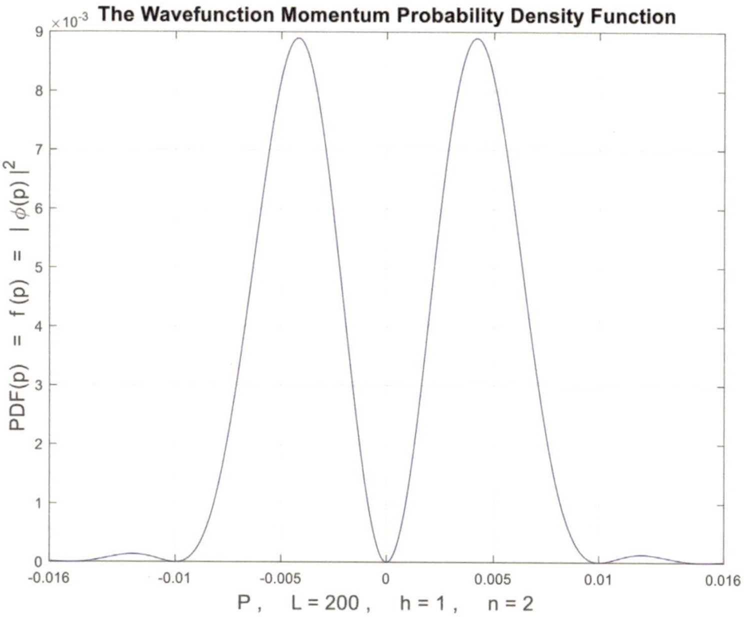

Figure 6.
The graph of the PDF *of the wavefunction momentum probability distribution as a function of the random variable* P *for* n = 2.

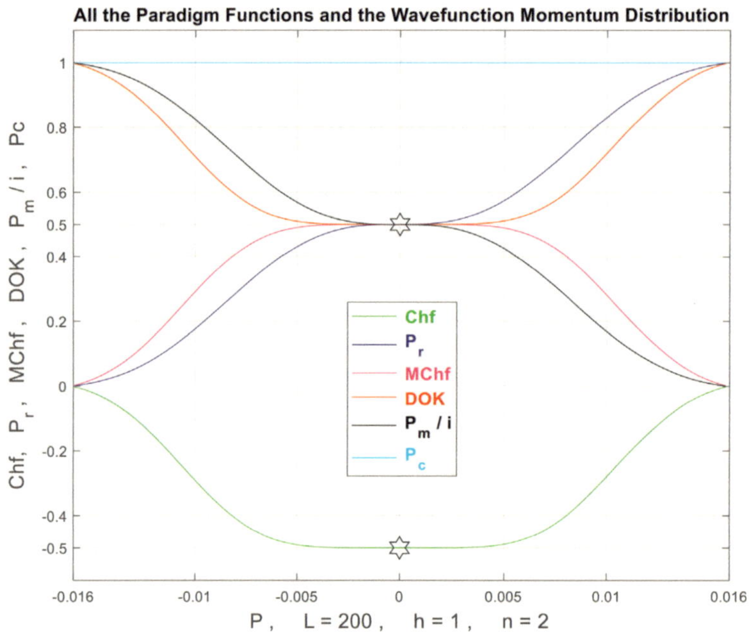

Figure 7.
The graphs of all the CPP *parameters as functions of the random variable* P *for the wavefunction momentum probability distribution for* n = 2.

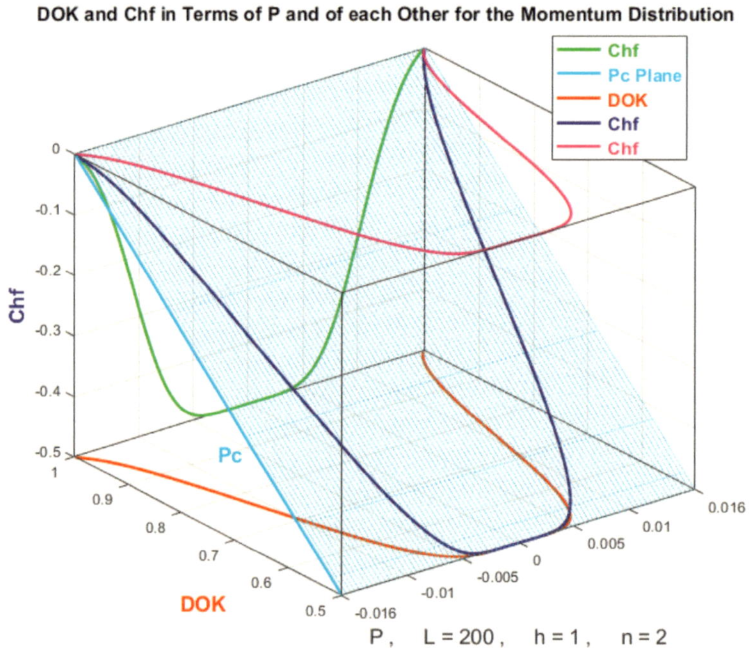

Figure 8.
The graphs of DOK *and* Chf *and the deterministic probability* Pc *in terms of* P *and of each other for the wavefunction momentum probability distribution for* n = 2.

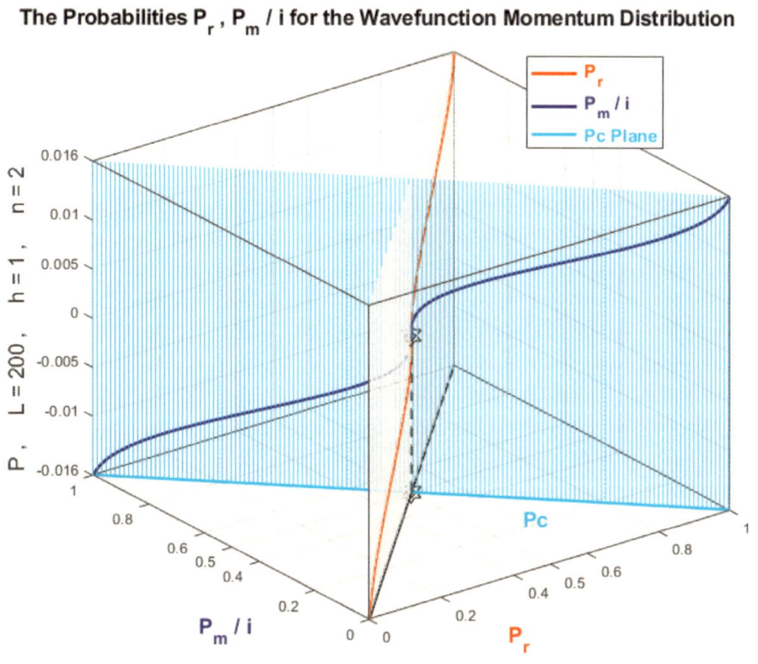

Figure 9.
The graphs of P_r *and* P_m/i *and* Pc *in terms of* P *and of each other for the wavefunction momentum probability distribution for* n = 2.

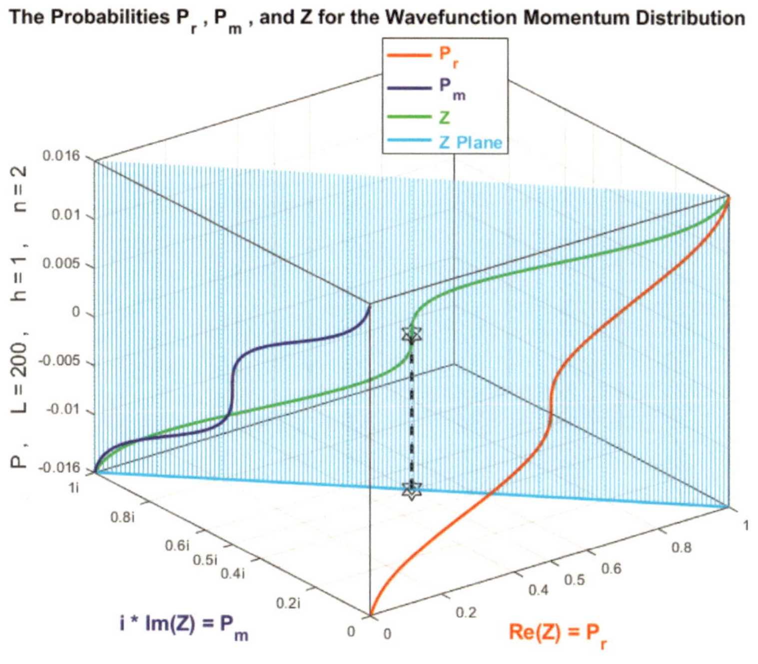

Figure 10.
The graphs of the probabilities P_r *and* P_m *and* Z *in terms of* P *for the wavefunction momentum probability distribution for* n = 2.

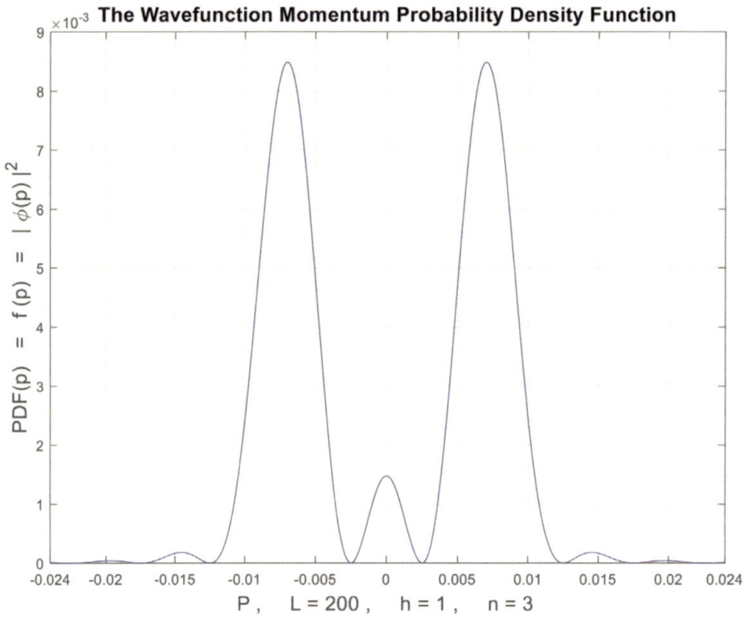

Figure 11.
The graph of the PDF *of the wavefunction momentum probability distribution as a function of the random variable* P *for* n = 3.

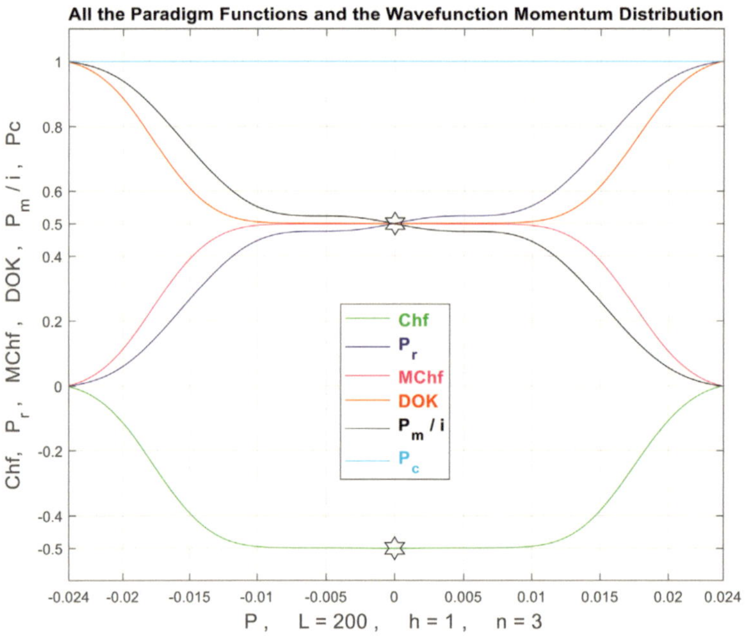

Figure 12.
The graphs of all the CPP *parameters as functions of the random variable* P *for the wavefunction momentum probability distribution for* n = 3.

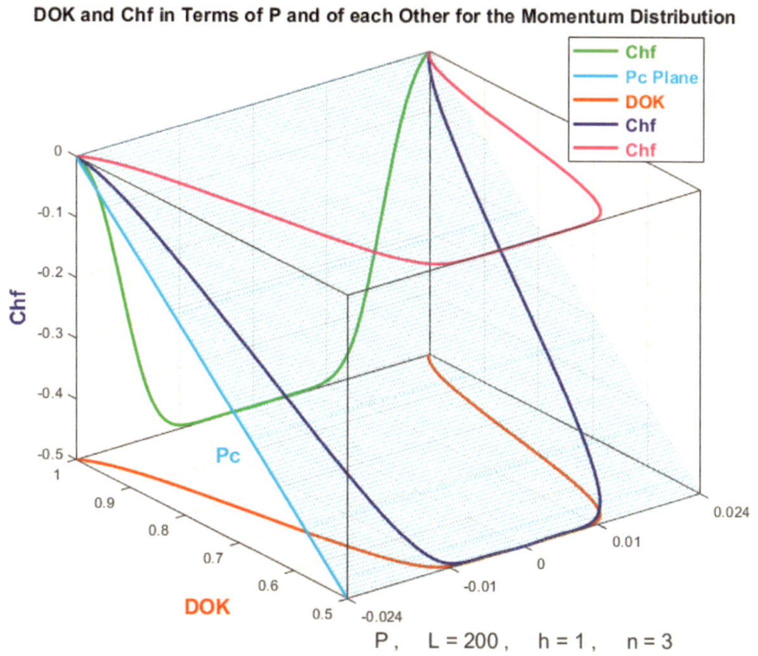

Figure 13.
The graphs of DOK *and* Chf *and the deterministic probability* Pc *in terms of* P *and of each other for the wavefunction momentum probability distribution for* n = 3.

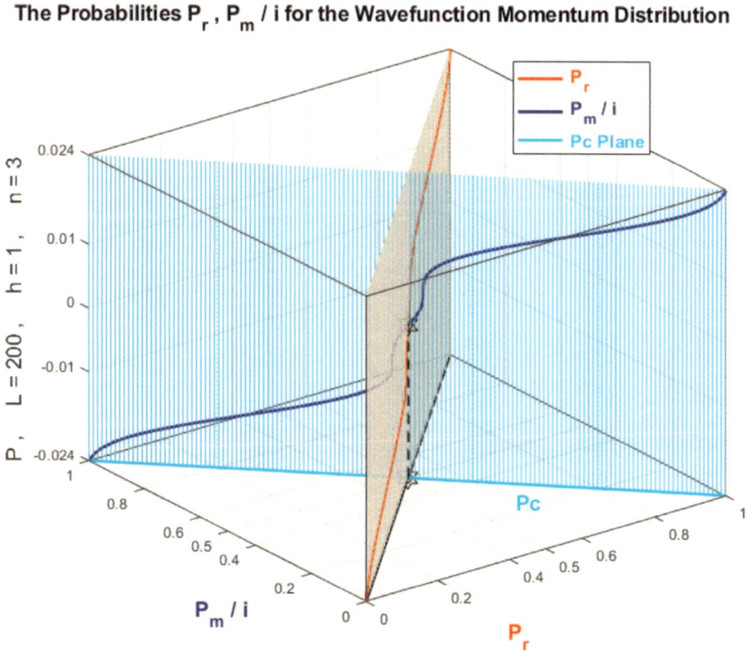

Figure 14.
The graphs of P_r *and* P_m/i *and* Pc *in terms of* P *and of each other for the wavefunction momentum probability distribution for* n = 3.

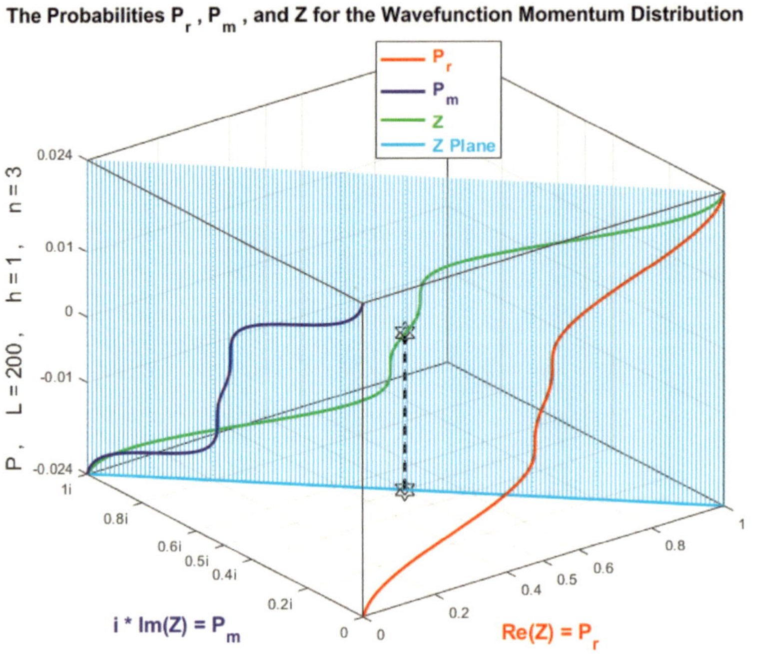

Figure 15.
The graphs of the probabilities P_r *and* P_m *and* Z *in terms of* P *for the wavefunction momentum probability distribution for* n = *3*.

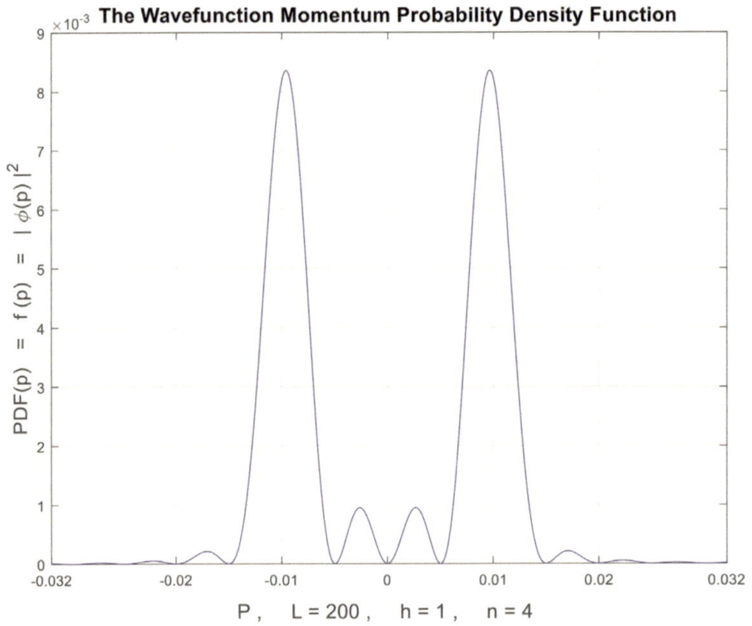

Figure 16.
The graph of the PDF *of the wavefunction momentum probability distribution as a function of the random variable* P *for* n = *4*.

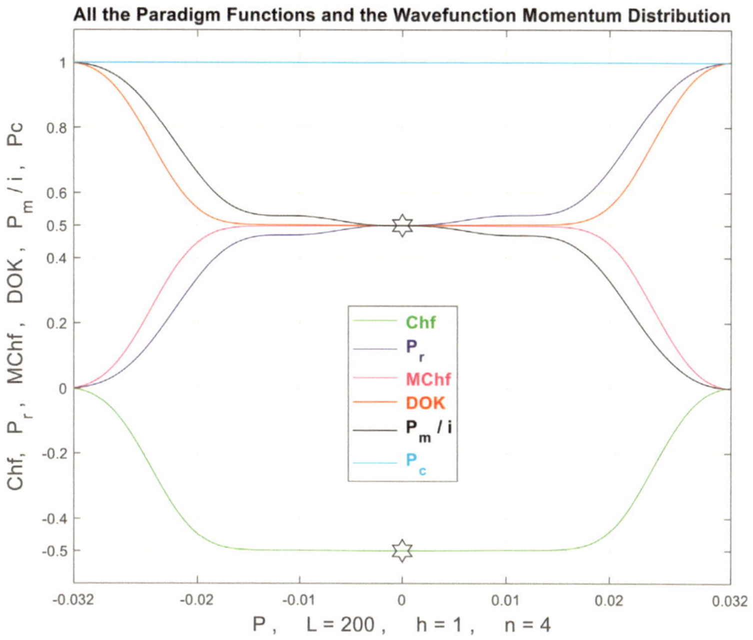

Figure 17.
The graphs of all the CPP *parameters as functions of the random variable* P *for the wavefunction momentum probability distribution for* n = 4.

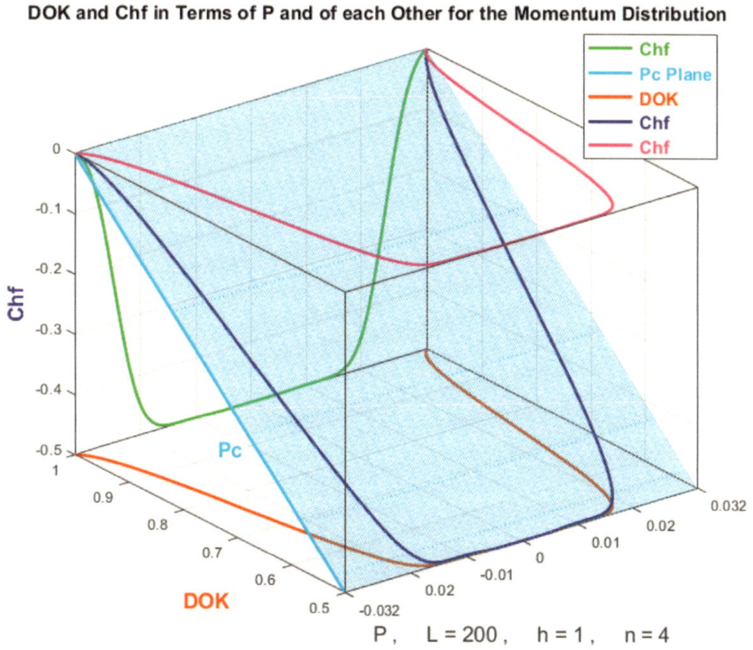

Figure 18.
The graphs of DOK *and* Chf *and the deterministic probability* Pc *in terms of* P *and of each other for the wavefunction momentum probability distribution for* n = 4.

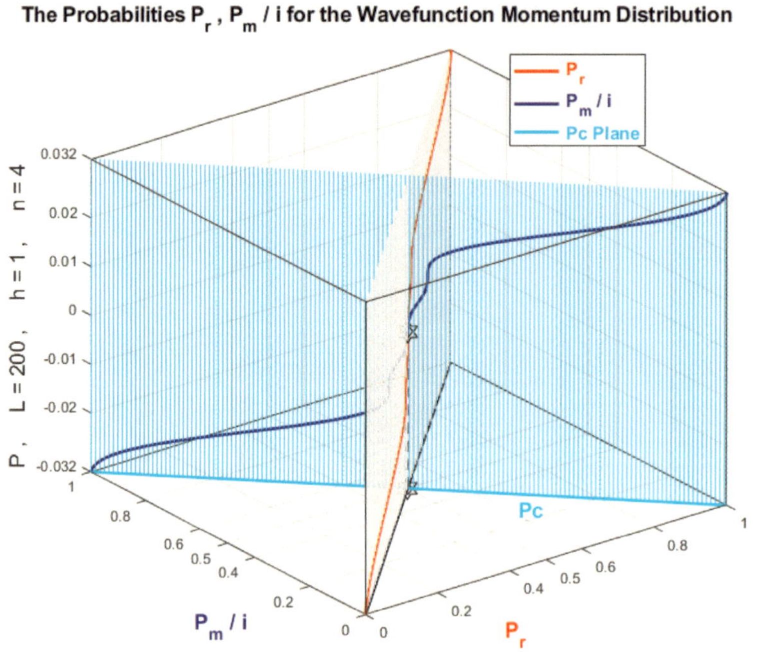

Figure 19.
The graphs of P_r *and* P_m/i *and* Pc *in terms of* P *and of each other for the wavefunction momentum probability distribution for* n = 4.

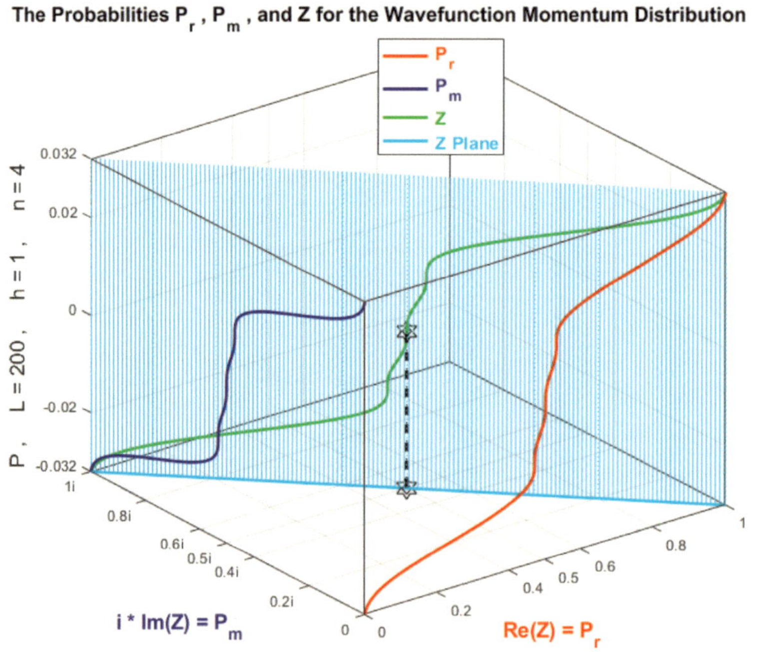

Figure 20.
The graphs of the probabilities P_r *and* P_m *and* Z *in terms of* P *for the wavefunction momentum probability distribution for* n = 4.

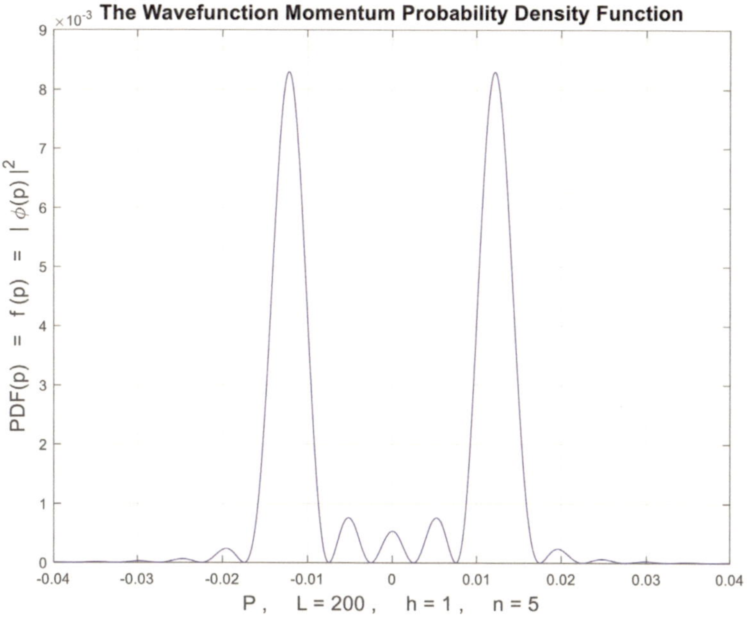

Figure 21.
The graph of the PDF *of the wavefunction momentum probability distribution as a function of the random variable* P *for* n = 5.

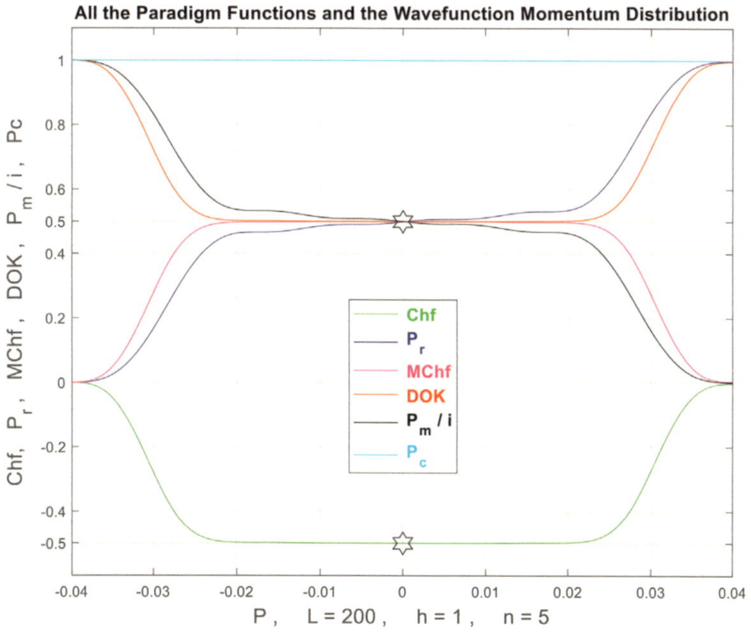

Figure 22.
The graphs of all the CPP *parameters as functions of the random variable* P *for the wavefunction momentum probability distribution for* n = 5.

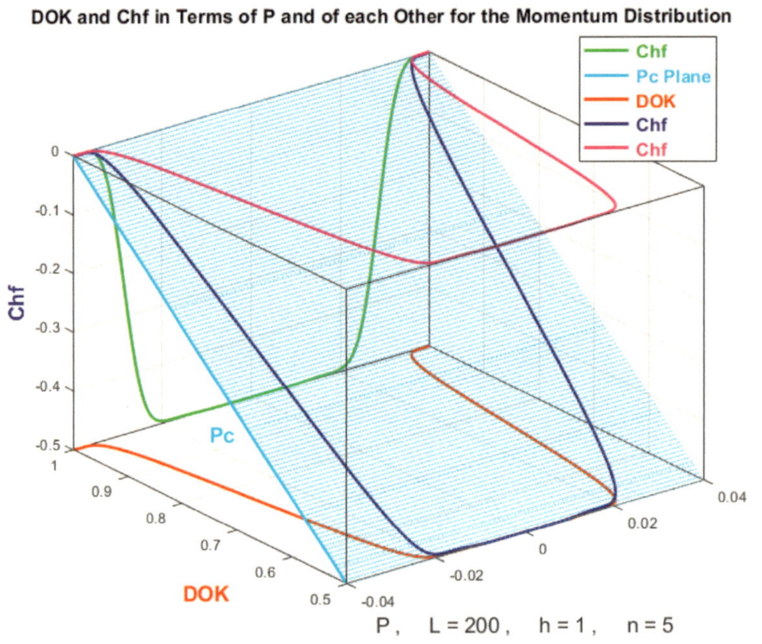

Figure 23.
The graphs of DOK *and* Chf *and the deterministic probability* Pc *in terms of* P *and of each other for the wavefunction momentum probability distribution for* n = 5.

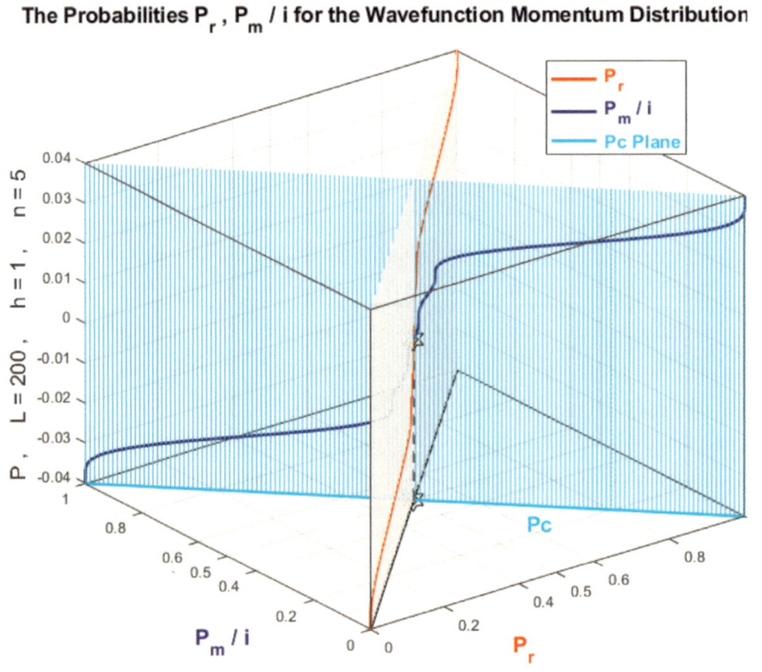

Figure 24.
The graphs of P_r *and* P_m/i *and* Pc *in terms of* P *and of each other for the wavefunction momentum probability distribution for* n = 5.

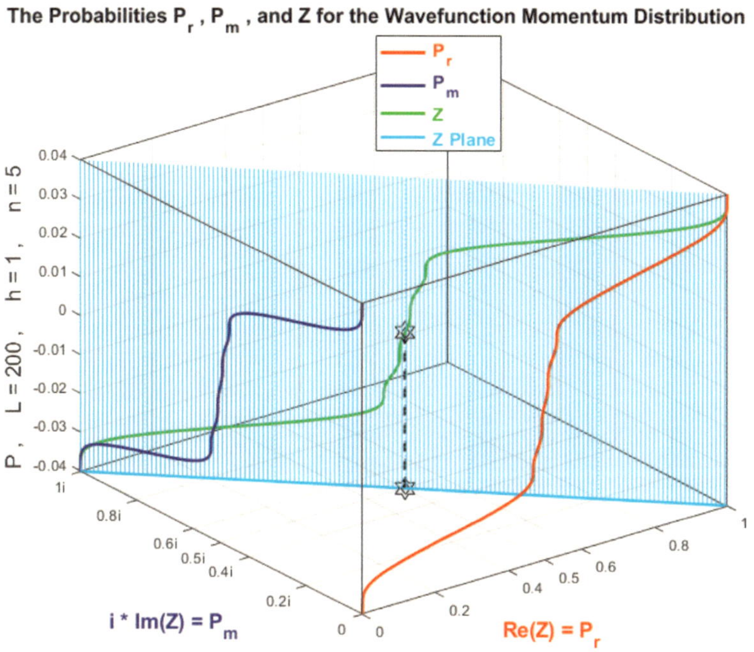

Figure 25.
The graphs of the probabilities P_r *and* P_m *and* Z *in terms of* P *for the wavefunction momentum probability distribution for* n = 5.

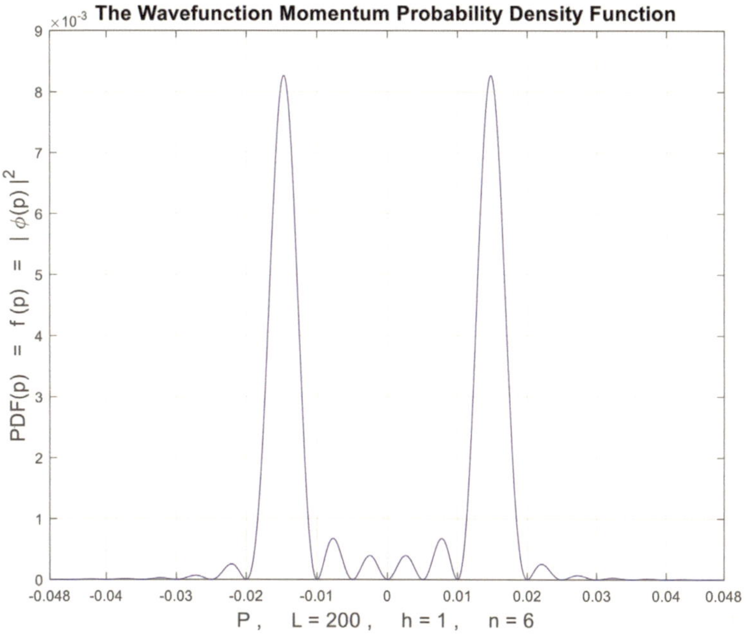

Figure 26.
The graph of the PDF *of the wavefunction momentum probability distribution as a function of the random variable* P *for* n = 6.

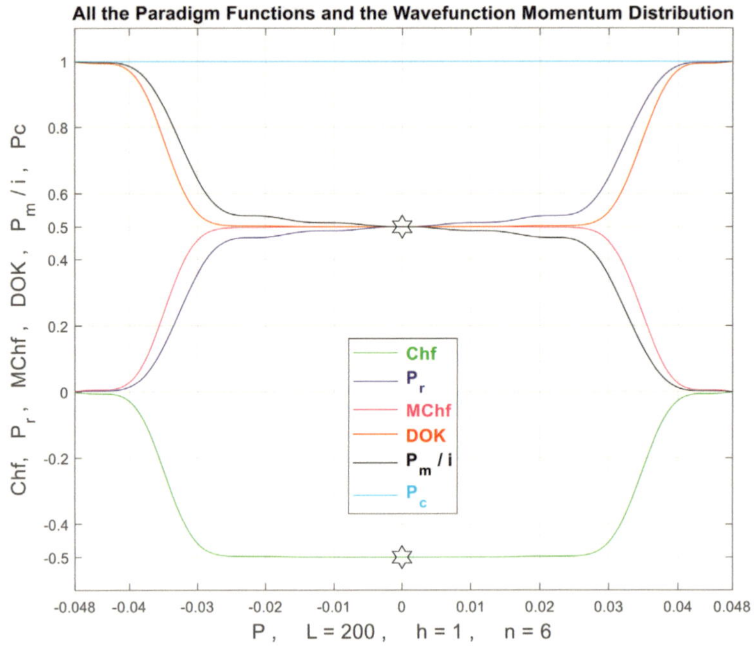

Figure 27.
The graphs of all the CPP *parameters as functions of the random variable* P *for the wavefunction momentum probability distribution for* n = 6.

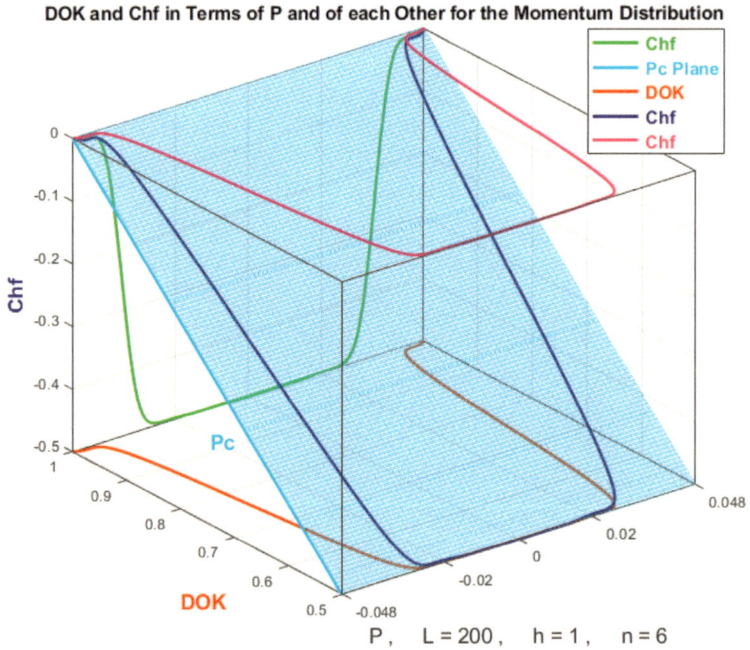

Figure 28.
The graphs of DOK *and* Chf *and the deterministic probability* Pc *in terms of* P *and of each other for the wavefunction momentum probability distribution for* n = 6.

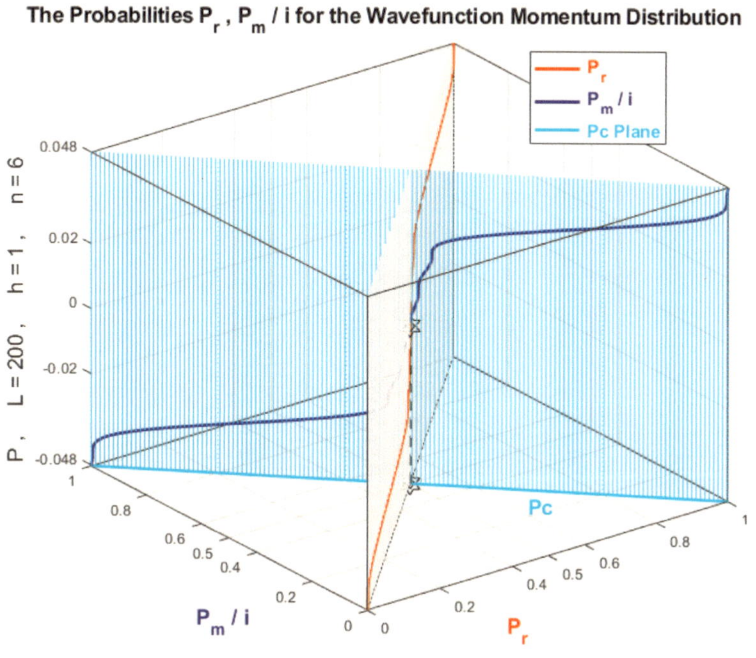

Figure 29.
The graphs of P_r *and* P_m/i *and* Pc *in terms of* P *and of each other for the wavefunction momentum probability distribution for* n = 6.

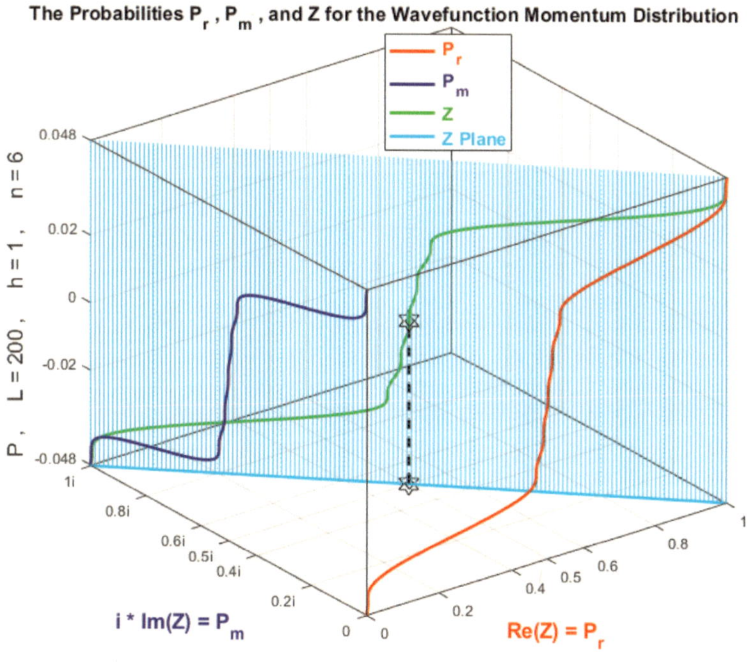

Figure 30.
The graphs of the probabilities P_r *and* P_m *and* Z *in terms of* P *for the wavefunction momentum probability distribution for* n = 6.

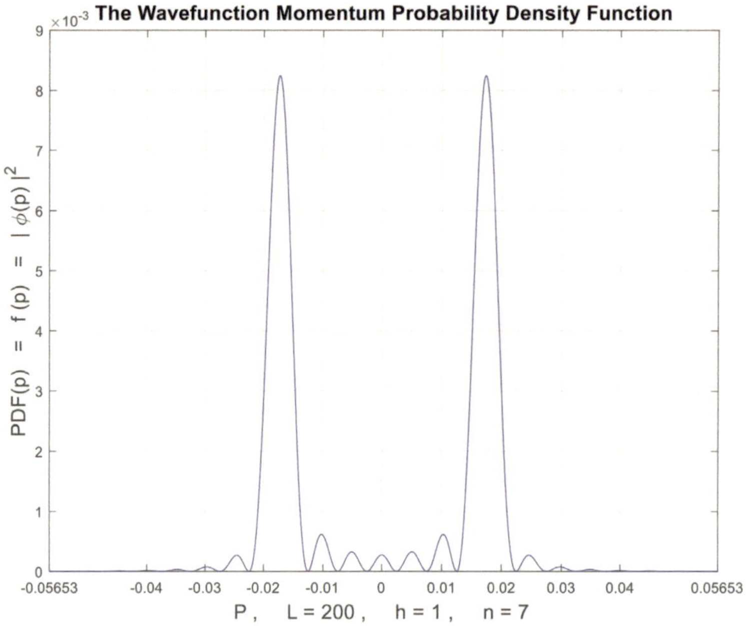

Figure 31.
The graph of the PDF *of the wavefunction momentum probability distribution as a function of the random variable* P *for* n = 7.

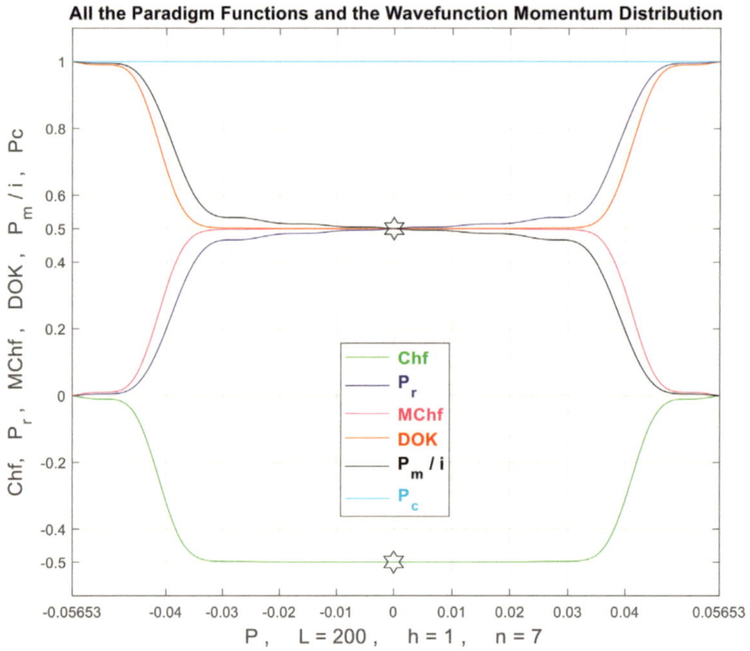

Figure 32.
The graphs of all the CPP *parameters as functions of the random variable* P *for the wavefunction momentum probability distribution for* n = 7.

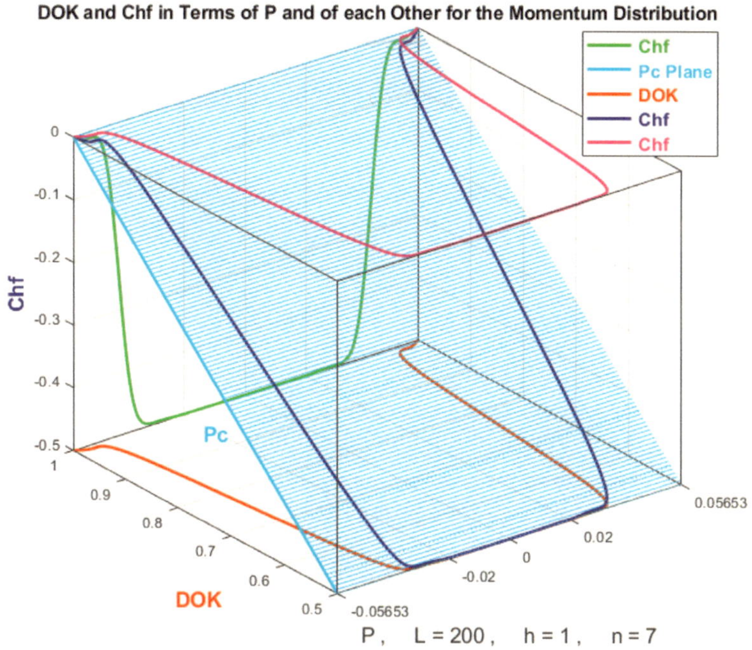

Figure 33.
The graphs of DOK *and* Chf *and the deterministic probability* Pc *in terms of* P *and of each other for the wavefunction momentum probability distribution for* n = 7.

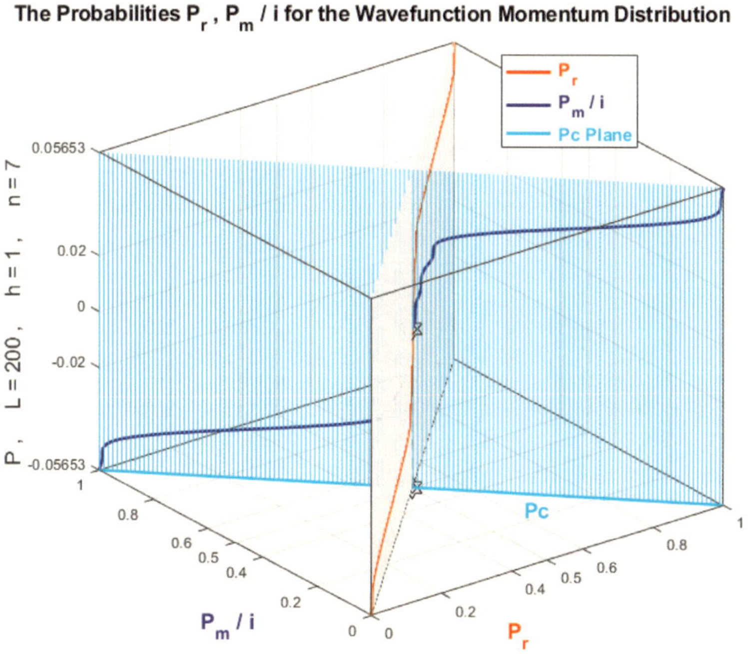

Figure 34.
The graphs of P_r *and* P_m/i *and* Pc *in terms of* P *and of each other for the wavefunction momentum probability distribution for* n = 7.

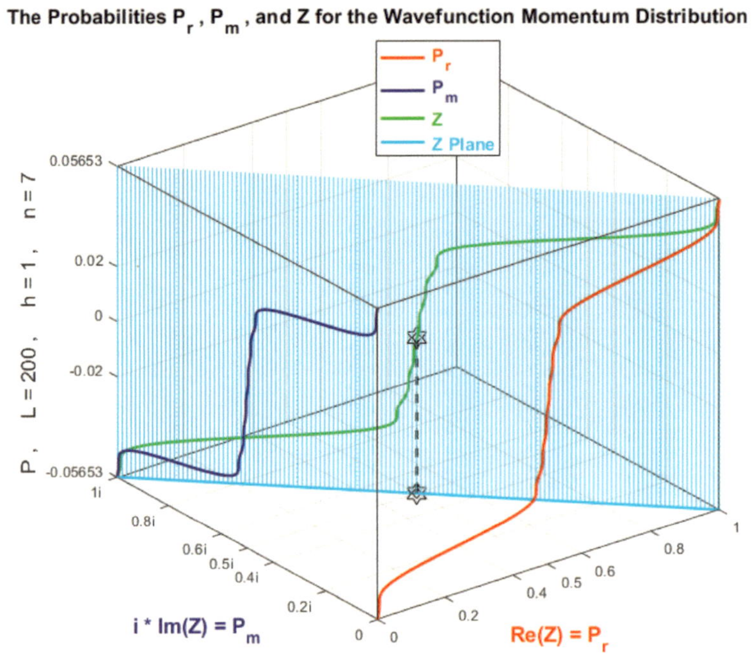

Figure 35.
The graphs of the probabilities P_r *and* P_m *and* Z *in terms of* P *for the wavefunction momentum probability distribution for* n = 7.

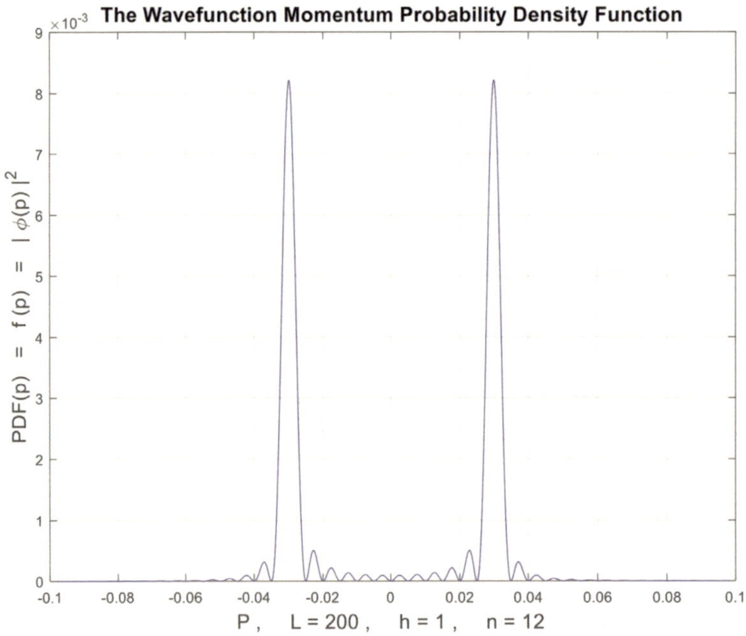

Figure 36.
The graph of the PDF *of the wavefunction momentum probability distribution as a function of the random variable* P *for* n = 12.

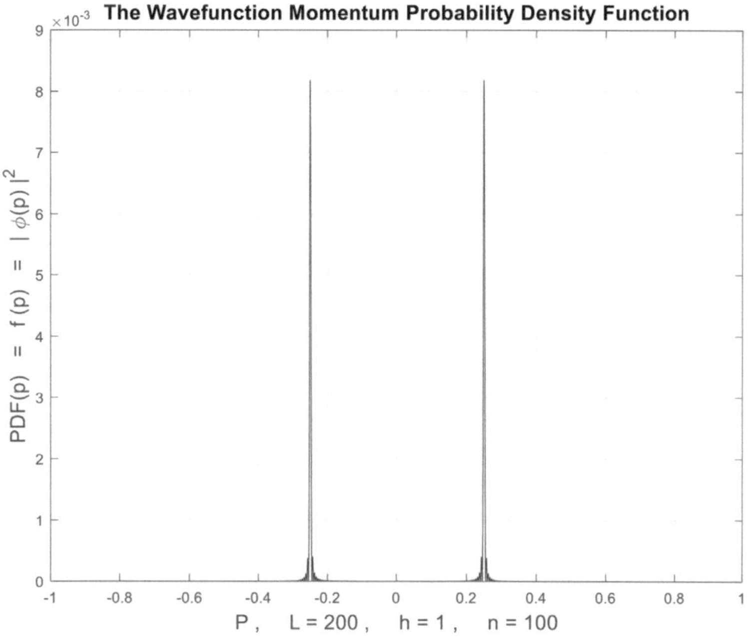

Figure 37.
The graph of the PDF *of the wavefunction momentum probability distribution as a function of the random variable* P *for* n = 100.

In the cubes (**Figures 3, 8, 13, 18, 23, 28,** and **33**), the simulation of *DOK* and *Chf* as functions of each other and the random variable *P* for the infinite potential well problem wavefunction momentum probability distribution can be seen. The thick line in cyan is the projection of the plane $Pc^2(P) = DOK(P) - Chf(P) = 1 = Pc(P)$ on the plane $P = L_b$ = lower bound of *P*. This thick line starts at the point ($DOK = 1$, $Chf = 0$) when $P = L_b$, reaches the point ($DOK = 0.5$, $Chf = -0.5$) when $P = 0$, and returns at the end to ($DOK = 1$, $Chf = 0$) when $P = U_b$ = upper bound of *P*. The other curves are the graphs of $DOK(P)$ (red) and $Chf(P)$ (green, blue, pink) in different simulation planes. Notice that they all have a minimum at the point ($DOK = 0.5$, $Chf = -0.5$, $P = 0$). The last simulation point corresponds to ($DOK = 1$, $Chf = 0$, $P = U_b$).

In the cubes (**Figures 4, 9, 14, 19, 24, 29,** and **34**), we can notice the simulation of the real probability $P_r(P)$ in $\mathcal{R}$ and its complementary real probability $P_m(P)/i$ in $\mathcal{R}$ also in terms of the random variable *P* for the infinite potential well problem wavefunction momentum probability distribution. The thick line in cyan is the projection of the plane $Pc^2(P) = P_r(P) + P_m(P)/i = 1 = Pc(P)$ on the plane $P = L_b$ = lower bound of *P*. This thick line starts at the point ($P_r = 0$, $P_m/i = 1$) and ends at the point ($P_r = 1$, $P_m/i = 0$). The red curve represents $P_r(P)$ in the plane $P_r(P) = P_m(P)/i$ in light gray. This curve starts at the point ($P_r = 0$, $P_m/i = 1$, $P = L_b$ = lower bound of *P*), reaches the point ($P_r = 0.5$, $P_m/i = 0.5$, $P = 0$), and gets at the end to ($P_r = 1$, $P_m/i = 0$, $P = U_b$ = upper bound of *P*). The blue curve represents $P_m(P)/i$ in the plane in cyan $P_r(P) + P_m(P)/i = 1 = Pc(P)$. Notice the importance of the point which is the intersection of the red and blue curves at $P = 0$ and when $P_r(P) = P_m(P)/i = 0.5$.

In the cubes (**Figures 5, 10, 15, 20, 25, 30**, and **35**), we can notice the simulation of the complex probability $Z(P)$ in $\mathcal{C} = \mathcal{R} + \mathcal{M}$ as a function of the real probability $P_r(P)$ = Re(Z) in $\mathcal{R}$ and of its complementary imaginary probability $P_m(P) = i \times \text{Im}(Z)$ in $\mathcal{M}$, and this in terms of the random variable P for the infinite potential well problem wavefunction momentum probability distribution. The red curve represents $P_r(P)$ in the plane $P_m(P) = 0$ and the blue curve represents $P_m(P)$ in the plane $P_r(P) = 0$. The green curve represents the complex probability $Z(P) = P_r(P) + P_m(P) = \text{Re}(Z) + i \times \text{Im}(Z)$ in the plane $P_r(P) = iP_m(P) + 1$ or $Z(P)$ plane in cyan. The curve of $Z(P)$ starts at the point ($P_r = 0, P_m = i, P = L_b$ = lower bound of P) and ends at the point ($P_r = 1, P_m = 0$, $P = U_b$ = upper bound of P). The thick line in cyan is $P_r(P = L_b) = iP_m(P = L_b) + 1$ and it is the projection of the $Z(P)$ curve on the complex probability plane whose equation is $P = L_b$. This projected thick line starts at the point ($P_r = 0, P_m = i, P = L_b$) and ends at the point ($P_r = 1, P_m = 0, P = L_b$). Notice the importance of the point corresponding to $P = 0$ and $Z = 0.5 + 0.5i$ when $P_r = 0.5$ and $P_m = 0.5i$.

1.1.3 The characteristics of the momentum probability distribution

In quantum mechanics, the average, or expectation value of the momentum of a particle is given by: $\langle p \rangle = \int_{-\infty}^{+\infty} p|\phi(p)|^2 dp = \int_{-\infty}^{+\infty} p \frac{L}{\pi\hbar}\left(\frac{n\pi}{n\pi+pL/\hbar}\right)^2 \text{sinc}^2\left[\frac{1}{2}(n\pi - pL/\hbar)\right]dp.$

For the steady state particle in a box, it can be shown that the average momentum is always $\langle p \rangle = 0$ regardless of the state of the particle. In the probability set and universe $\mathcal{R}$, we have:

$$\langle p \rangle_R = \langle p \rangle = 0$$

The variance in the momentum is a measure of the uncertainty in momentum of the particle, so in the probability set and universe $\mathcal{R}$, we have:

$$\text{Var}_{p,R} = \text{Var}(p) = \langle p^2 \rangle_R - \langle p \rangle_R^2 = \int_{-\infty}^{+\infty} p^2|\phi(p)|^2 dp - 0$$

$$= \int_{-\infty}^{+\infty} p^2\left\{\frac{L}{\pi\hbar}\left(\frac{n\pi}{n\pi+pL/\hbar}\right)^2 \text{sinc}^2\left[\frac{1}{2}(n\pi - pL/\hbar)\right]\right\}dp = \left(\frac{\hbar n\pi}{L}\right)^2$$

In the probability set and universe $\mathcal{M}$, we have:

$$\langle p \rangle_M = \int_{-\infty}^{+\infty} p\left\{i\left[1 - |\phi(p)|^2\right]\right\}dp = i\int_{-\infty}^{+\infty} p\left\{1 - \frac{L}{\pi\hbar}\left(\frac{n\pi}{n\pi+pL/\hbar}\right)^2 \text{sinc}^2\left[\frac{1}{2}(n\pi - pL/\hbar)\right]\right\}dp$$

$$= i\left\{\int_{-\infty}^{+\infty} pdp - \int_{-\infty}^{+\infty} p\left\{\frac{L}{\pi\hbar}\left(\frac{n\pi}{n\pi+pL/\hbar}\right)^2 \text{sinc}^2\left[\frac{1}{2}(n\pi - pL/\hbar)\right]\right\}dp\right\}$$

$$= i\left\{\left[\frac{p^2}{2}\right]_{-\infty}^{+\infty} - \langle p \rangle_R\right\} = i\left\{\left[\frac{p^2}{2}\right]_{-U_b}^{U_b} - \langle p \rangle_R\right\} = i\{0 - 0\} = 0$$

$$\begin{aligned}
\mathrm{Var}_{p,M} &= \langle p^2\rangle_M - \langle p\rangle_M^2 \\
&= \int_{-\infty}^{+\infty} p^2\left\{i\left[1-|\phi(p)|^2\right]\right\}dp - 0 \\
&= i\int_{-\infty}^{+\infty} p^2\left\{1-\frac{L}{\pi\hbar}\left(\frac{n\pi}{n\pi+pL/\hbar}\right)^2\mathrm{sinc}^2\left[\frac{1}{2}(n\pi-pL/\hbar)\right]\right\}dp \\
&= i\left\{\int_{-\infty}^{+\infty} p^2dp - \int_{-\infty}^{+\infty} p^2\left\{\frac{L}{\pi\hbar}\left(\frac{n\pi}{n\pi+pL/\hbar}\right)^2\mathrm{sinc}^2\left[\frac{1}{2}(n\pi-pL/\hbar)\right]\right\}dp\right\} \\
&= i\left\{\int_{-\infty}^{+\infty} p^2dp - \mathrm{Var}_{p,R}\right\} = i\left\{\left[\frac{p^3}{3}\right]_{-\infty}^{+\infty} - \mathrm{Var}_{p,R}\right\} \to i\left\{+\infty - \left(\frac{\hbar n\pi}{L}\right)^2\right\} \\
&\to +\infty
\end{aligned}$$

In the probability set and the universe $\mathcal{C} = \mathcal{R} + \mathcal{M}$, we have from *CPP*:

$$\begin{aligned}
\langle p\rangle_C &= \int_{-\infty}^{+\infty} p[z(p)]dp = \int_{-\infty}^{+\infty} p\left\{|\phi(p)|^2 + i\left[1-|\phi(p)|^2\right]\right\}dp \\
&= \int_{-\infty}^{+\infty} p|\phi(p)|^2dp + \int_{-\infty}^{+\infty} pi\left[1-|\phi(p)|^2\right]dp \\
&= \langle p\rangle_R + \langle p\rangle_M = 0 + i(0) = 0
\end{aligned}$$

$$\begin{aligned}
\mathrm{Var}_{p,C} &= \langle p^2\rangle_C - \langle p\rangle_C^2 = \left[\int_{-\infty}^{+\infty} p^2[z(p)]dp\right] - \left[\langle p\rangle_R + \langle p\rangle_M\right]^2 \\
&= \left[\int_{-\infty}^{+\infty} p^2\left\{|\phi(p)|^2 + i\left[1-|\phi(p)|^2\right]\right\}dp\right] - \left[\langle p\rangle_R + \langle p\rangle_M\right]^2 \\
&= \left[\int_{-\infty}^{+\infty} p^2|\phi(p)|^2dp + \int_{-\infty}^{+\infty} p^2i\left[1-|\phi(p)|^2\right]dp\right] - \left[\langle p\rangle_R + \langle p\rangle_M\right]^2 \\
&= \left[\langle p^2\rangle_R + \langle p^2\rangle_M\right] - \left[\langle p\rangle_R + \langle p\rangle_M\right]^2 \\
&= \left[\langle p^2\rangle_R + \langle p^2\rangle_M\right] - \left[\langle p\rangle_R^2 + \langle p\rangle_M^2 + 2\langle p\rangle_R\langle p\rangle_M\right] \\
&= \left[\langle p^2\rangle_R - \langle p\rangle_R^2\right] + \left[\langle p^2\rangle_M - \langle p\rangle_M^2\right] - 2\langle p\rangle_R\langle p\rangle_M \\
&= \mathrm{Var}_{p,R} + \mathrm{Var}_{p,M} - 2\langle p\rangle_R\langle p\rangle_M \\
&\to \left(\frac{\hbar n\pi}{L}\right)^2 + \infty - 2(0)(0) \\
&\to +\infty
\end{aligned}$$

Momentum distribution characteristics	$L = 200, h = 1, n = 1$
$\langle p \rangle_R$	0
$\text{Var}_{p,R}$	6.2500e−06
$\langle p \rangle_M$	0
$\text{Var}_{p,M}$	$+\infty$
$\langle p \rangle_C = \langle p \rangle_R + \langle p \rangle_M$	$0 + i(0)$
$\text{Var}_{p,C} = \text{Var}_{p,R} + \text{Var}_{p,M} - 2\langle p \rangle_R \langle p \rangle_M$	$+\infty$

Table 1.
The momentum distribution characteristics for L = 200, h = 1, and n = 1.

Momentum distribution characteristics	$L = 200, h = 1, n = 2$
$\langle p \rangle_R$	0
$\text{Var}_{p,R}$	2.500e−05
$\langle p \rangle_M$	0
$\text{Var}_{p,M}$	$+\infty$
$\langle p \rangle_C = \langle p \rangle_R + \langle p \rangle_M$	$0 + i(0)$
$\text{Var}_{p,C} = \text{Var}_{p,R} + \text{Var}_{p,M} - 2\langle p \rangle_R \langle p \rangle_M$	$+\infty$

Table 2.
The momentum distribution characteristics for L = 200, h = 1, and n = 2.

Momentum distribution characteristics	$L = 200, h = 1, n = 8$
$\langle p \rangle_R$	0
$\text{Var}_{p,R}$	4.0000e−04
$\langle p \rangle_M$	0
$\text{Var}_{p,M}$	$+\infty$
$\langle p \rangle_C = \langle p \rangle_R + \langle p \rangle_M$	$0 + i(0)$
$\text{Var}_{p,C} = \text{Var}_{p,R} + \text{Var}_{p,M} - 2\langle p \rangle_R \langle p \rangle_M$	$+\infty$

Table 3.
The momentum distribution characteristics for L = 200, h = 1, and n = 8.

The following tables (**Tables 1–4**) compute the momentum distribution characteristics for $L = 200$, $h = 1$, and $n = 1,2,8,10000$.

For $n \gg 1$ (large n) we get: $\text{Var}_{p,R} = \left(\frac{\hbar n \pi}{L}\right)^2 \to +\infty$.

2. Heisenberg uncertainty principle in $\mathcal{R}$, $\mathcal{M}$, and $\mathcal{C}$

The uncertainties in the probability set and universe $\mathcal{R}$ in position and momentum (Δx_R and Δp_R) are defined as being equal to the square root of their respective variances in $\mathcal{R}$, so that:

Momentum distribution characteristics	$L = 200, h = 1, n = 10{,}000$
$\langle p \rangle_R$	0
$\text{Var}_{p,R}$	625
$\langle p \rangle_M$	0
$\text{Var}_{p,M}$	$+\infty$
$\langle p \rangle_C = \langle p \rangle_R + \langle p \rangle_M$	$0 + i(0)$
$\text{Var}_{p,C} = \text{Var}_{p,R} + \text{Var}_{p,M} - 2\langle p \rangle_R \langle p \rangle_M$	$+\infty$

Table 4.
The momentum distribution characteristics for L = 200, h = 1, and n = 10,000.

$$\Delta x_R \times \Delta p_R = \sqrt{\text{Var}_{x,R}} \times \sqrt{\text{Var}_{p,R}} = \sqrt{\frac{L^2}{12}\left(1 - \frac{6}{n^2\pi^2}\right)} \times \sqrt{\frac{\hbar^2 n^2 \pi^2}{L^2}} = \frac{\hbar}{2}\sqrt{\frac{n^2\pi^2}{3} - 2}$$

This product increases with increasing n, having a minimum value for $n = 1$. The value of this product for $n = 1$ is about equal to 0.568 $\hbar$ which obeys the Heisenberg uncertainty principle, which states that:

$$\Delta x \times \Delta p \geq \frac{\hbar}{2} \Leftrightarrow \forall n \geq 1 : \Delta x_R \times \Delta p_R \geq \frac{\hbar}{2}$$

The uncertainties in the probability set and universe $\mathcal{M}$ in position and momentum (Δx_M and Δp_M) are defined as being equal to the square root of their respective variances in $\mathcal{M}$, so that:

$$\Delta x_M \times \Delta p_M = \sqrt{\text{Var}_{x,M}} \times \sqrt{\text{Var}_{p,M}} \to \sqrt{i\left\{\frac{L^2}{12}\left[L - \left(1 - \frac{6}{n^2\pi^2}\right)\right]\right\}} \times \sqrt{+\infty} \to +\infty$$

$\Leftrightarrow \forall n \geq 1 : \Delta x_M \times \Delta p_M \geq \frac{\hbar}{2}$, in accordance with the Heisenberg uncertainty principle.

The uncertainties in the probability set and universe $\mathcal{C} = \mathcal{R} + \mathcal{M}$ in position and momentum (Δx_C and Δp_C) are defined as being equal to the square root of their respective variances in $\mathcal{C}$, so that:

$$\Delta x_C \times \Delta p_C = \sqrt{\text{Var}_{x,C}} \times \sqrt{\text{Var}_{p,C}}$$
$$\to \sqrt{\frac{L^2}{12}\left(1 - \frac{6}{n^2\pi^2}\right) + i\left\{\frac{L^2}{12}\left[L - \left(1 - \frac{6}{n^2\pi^2}\right)\right]\right\}} \times \sqrt{+\infty} \to +\infty$$

$\Leftrightarrow \forall n \geq 1 : \Delta x_C \times \Delta p_C \geq \frac{\hbar}{2}$, in accordance with the Heisenberg uncertainty principle.

Consequently, the Heisenberg uncertainty principle is verified in the universe $\mathcal{R}$, in the universe $\mathcal{M}$, and the complex universe $\mathcal{C}$.

3. The Wavefunction Entropies in $\mathcal{R}$,$\mathcal{M}$,and $\mathcal{C}$

Another measure of uncertainty in position is the information entropy of the probability distribution H_x which is the entropy in $\mathcal{R}$ and is equal to:

$$H_x = -\sum_{x=-\infty}^{x=+\infty} |\psi(x)|^2 Ln\left[|\psi(x)|^2 x_0\right] = -\sum_{x=x_c-\frac{L}{2}}^{x=x_c+\frac{L}{2}} |\psi(x)|^2 Ln\left[|\psi(x)|^2 x_0\right] = H_x^R = Ln\left(\frac{2L}{ex_0}\right)$$

where x_0 is an arbitrary reference length [1, 2]. Take $x_0 = 1$:

$$\Leftrightarrow H_x^R = -\sum_{x=x_c-\frac{L}{2}}^{x=x_c+\frac{L}{2}} |\psi(x)|^2 Ln\left[|\psi(x)|^2\right]$$
$$= Ln\left(\frac{2L}{e}\right) = Ln(2L) - Ln(e) = Ln(2L) - 1 = Ln(2 \times 200) - 1 = 4.991464547\ldots$$

$\Leftrightarrow \forall x : x_c - \frac{L}{2} \leq x \leq x_c + \frac{L}{2}$, we have : $d\left[H_x^R\right] \geq 0$, that means that H_x^R is a nondecreasing series with x and converging to $Ln\left(\frac{2L}{e}\right)$ and that also in $\mathcal{R}$, chaos and disorder are increasing with x.

The negative real entropy corresponding to H_x^R in $\mathcal{R}$ is $NegH_x^R$ and is the following:

$$NegH_x^R = -H_x^R = \sum_{x=-\infty}^{x=+\infty} |\psi(x)|^2 Ln\left[|\psi(x)|^2\right] = \sum_{x=x_c-\frac{L}{2}}^{x=x_c+\frac{L}{2}} |\psi(x)|^2 Ln\left[|\psi(x)|^2\right] = -Ln\left(\frac{2L}{e}\right)$$
$$= 1 - Ln(2L) = 1 - Ln(2 \times 200) = -4.991464547\ldots$$

$\Leftrightarrow \forall x : x_c - \frac{L}{2} \leq x \leq x_c + \frac{L}{2}$, we have : $d\left[NegH_x^R\right] \leq 0$, which means that $NegH_x^R$ is a nonincreasing series with x and converging to $-Ln\left(\frac{2L}{e}\right)$. Therefore, if H_x^R measures in $\mathcal{R}$ the amount of disorder, of uncertainty, of chaos, of ignorance, of unpredictability, and of information gain in a random system then since $NegH_x^R = -H_x^R$, that means the opposite of H_x^R, $NegH_x^R$ measures in $\mathcal{R}$ the amount of order, of certainty, of predictability, and of information loss in a stochastic system.

The complementary real entropy to H_x^R in $\mathcal{R}$ is $\overline{H}_x^R$ and is the following:

$$\overline{H}_x^R = -\sum_{x=-\infty}^{x=+\infty}\left[1 - |\psi(x)|^2\right] Ln\left[1 - |\psi(x)|^2\right] = -\sum_{x=x_c-\frac{L}{2}}^{x=x_c+\frac{L}{2}}\left[1 - |\psi(x)|^2\right] Ln\left[1 - |\psi(x)|^2\right] = 1$$

In the complementary real probability set to $\mathcal{R}$, we denote the corresponding real entropy by $\overline{H}_x^R$.

The meaning of $\overline{H}_x^R$ is the following: it is the real entropy in the real set $\mathcal{R}$ and which is related to the complementary real probability $P_m/i = 1 - P_r$.

$\Leftrightarrow \forall x : x_c - \frac{L}{2} \leq x \leq x_c + \frac{L}{2}$, we have : $d\left[\overline{H}_x^R\right] \geq 0$, that means that $\overline{H}_x^R$ is a nondecreasing series with x and converging to 1 and that also means that in the complementary real probability set to $\mathcal{R}$, chaos and disorder are increasing with x.

In the complementary imaginary probability set $\mathcal{M}$ to the set $\mathcal{R}$, we denote the corresponding imaginary entropy by H_x^M. The meaning of H_x^M is the following: it is the imaginary entropy in the imaginary set $\mathcal{M}$ and which is related to the complementary imaginary probability $P_m = i(1 - P_r)$. The complementary entropy to H_x^R in $\mathcal{M}$ is H_x^M and is computed as follows:

$$
\begin{aligned}
H_x^M &= -\sum_{x=-\infty}^{x=+\infty} i\left[1 - |\psi(x)|^2\right] Ln\left\{i\left[1 - |\psi(x)|^2\right]\right\} \\
&= -\sum_{x=x_c-\frac{L}{2}}^{x=x_c+\frac{L}{2}} i\left[1 - |\psi(x)|^2\right] Ln\left\{i\left[1 - |\psi(x)|^2\right]\right\} \\
&= -\sum_{x=x_c-\frac{L}{2}}^{x=x_c+\frac{L}{2}} i\left[1 - |\psi(x)|^2\right]\left\{Lni + Ln\left[1 - |\psi(x)|^2\right]\right\} \\
&= -\sum_{x=x_c-\frac{L}{2}}^{x=x_c+\frac{L}{2}} i\left\{Lni + Ln\left[1 - |\psi(x)|^2\right] - \left[|\psi(x)|^2\right] Lni - \left[|\psi(x)|^2\right] Ln\left[1 - |\psi(x)|^2\right]\right\} \\
&= -\sum_{x=x_c-\frac{L}{2}}^{x=x_c+\frac{L}{2}} iLni + iLn\left[1 - |\psi(x)|^2\right] - i\left[|\psi(x)|^2\right] Lni - i\left[|\psi(x)|^2\right] Ln\left[1 - |\psi(x)|^2\right] \\
&= -\sum_{x=x_c-\frac{L}{2}}^{x=x_c+\frac{L}{2}} iLni\left[1 - |\psi(x)|^2\right] + i\left[1 - |\psi(x)|^2\right] Ln\left[1 - |\psi(x)|^2\right] \\
&= -\sum_{x=x_c-\frac{L}{2}}^{x=x_c+\frac{L}{2}} iLni\left[1 - |\psi(x)|^2\right] - i\sum_{x=x_c-\frac{L}{2}}^{x=x_c+\frac{L}{2}} \left[1 - |\psi(x)|^2\right] Ln\left[1 - |\psi(x)|^2\right] \\
&= -\sum_{x=x_c-\frac{L}{2}}^{x=x_c+\frac{L}{2}} iLni\left[1 - |\psi(x)|^2\right] + i\overline{H}_x^R = -iLni\sum_{x=x_c-\frac{L}{2}}^{x=x_c+\frac{L}{2}} \left[1 - |\psi(x)|^2\right] + i\overline{H}_x^R \\
&= -iLni\left\{\sum_{x=x_c-\frac{L}{2}}^{x=x_c+\frac{L}{2}} 1 - \sum_{x=x_c-\frac{L}{2}}^{x=x_c+\frac{L}{2}} |\psi(x)|^2\right\} + i\overline{H}_x^R \\
&= -iLni\left\{\left[\left(x_c + \frac{L}{2}\right) - \left(x_c - \frac{L}{2}\right) + 1\right] - 1\right\} + i\overline{H}_x^R \quad \text{since} \quad \sum_{x=x_c-\frac{L}{2}}^{x=x_c+\frac{L}{2}} |\psi(x)|^2 = 1 \\
&= -(iLni)L + i\overline{H}_x^R
\end{aligned}
$$

From the properties of logarithms, we have: $\theta Ln x = Ln\left(x^{\theta}\right)$ then $iLni = Lni^{i}$.

Moreover, Leonhard Euler's formula for complex numbers gives: $e^{i\theta} = \cos\theta + i\sin\theta$.

Take $\theta = \pi/2 + 2k\pi \Leftrightarrow e^{i(\pi/2+2k\pi)} = \cos(\pi/2+2k\pi) + i\sin(\pi/2+2k\pi) = 0 + i(1) = i$, then:

$i^{i} = \left(e^{i(\pi/2+2k\pi)}\right)^{i} = e^{i^{2}(\pi/2+2k\pi)} = e^{-(\pi/2+2k\pi)}$ since $i^{2} = -1$, therefore:

$-iLni = -Lni^{i} = -Ln\left[e^{-(\pi/2+2k\pi)}\right] = \pi/2 + 2k\pi$ since $Ln[e] = 1$ and where k belongs to the set of integer numbers Z.

Consequently,

$$H_x^M = -(iLni)L + i\overline{H}_x^R = (\pi/2 + 2k\pi)L + i\overline{H}_x^R$$

That means that H_x^M is a complex number where:

the real part is: $\mathrm{Re}\left(H_x^M\right) = (\pi/2 + 2k\pi)L$, and the imaginary part is: $\mathrm{Im}\left(H_x^M\right) = \overline{H}_x^R$.

For $k = -1$ then

$\mathrm{Re}\left(H_x^M\right) = (-3\pi/2)L = -4.71238898L = -942.4777961\ldots$ for $L = 200$.

For $k = 0$ then $\mathrm{Re}\left(H_x^M\right) = (\pi/2)L = 1.570796327L = 314.1592654\ldots$ for $L = 200$.

For $k = 1$ then

$\mathrm{Re}\left(H_x^M\right) = (5\pi/2)L = 7.853981634L = 1570.796327\ldots$ for $L = 200$,

etc.

Finally, the entropy H_x^C in $\mathcal{C} = \mathcal{R} + \mathcal{M}$ is the following:

$$\begin{aligned} H_x^C &= -\sum_{x=x_c-\frac{L}{2}}^{x=x_c+\frac{L}{2}} Pc(x)Ln[Pc(x)] \\ &= -\sum_{x=x_c-\frac{L}{2}}^{x=x_c+\frac{L}{2}} 1 \times Ln[1] = -\sum_{x=x_c-\frac{L}{2}}^{x=x_c+\frac{L}{2}} (1 \times 0) = 0 \\ &= H_x^R + NegH_x^R \end{aligned}$$

$\Leftrightarrow \forall x : x_c - \frac{L}{2} \le x \le x_c + \frac{L}{2}$, we have: $d\left[H_x^C\right] = 0$, that means that H_x^C is a constant series with x and is always equal to 0. That means also and most importantly, for the wavefunction position distribution and in the probability set and universe $\mathcal{C} = \mathcal{R} + \mathcal{M}$, we have complete order, no chaos, no ignorance, no uncertainty, no disorder, no randomness, no information loss or gain but a conservation of information, and no unpredictability since all measurements are completely and perfectly deterministic ($Pc(x) = 1$ and $H_x^C = 0$).

Similarly, we can determine another measure of uncertainty in momentum which is the information entropy of the probability distribution H_p and which is [1, 2]:

$$H_p = -\sum_{p=-\infty}^{p=+\infty} |\phi(p)|^2 Ln\left[|\phi(p)|^2 p_0\right] = Ln\left(\frac{4\pi\hbar e^{2(1-\gamma)}}{Lp_0}\right) = \lim_{n\to+\infty} H_p(n)$$

Where γ is Euler's constant and is equal to: 0.577215664901532 ...

For $p_0 = 1$ we can compute all the defined entropies in $\mathcal{R}$, $\mathcal{M}$, and $\mathcal{C}$ and which are [1–30]:

$$H_p^R = -\sum_{p=-\infty}^{p=+\infty} |\phi(p)|^2 Ln\left[|\phi(p)|^2\right] = Ln\left(\frac{4\pi\hbar e^{2(1-\gamma)}}{L}\right) = \lim_{n\to+\infty} H_p(n)$$

$$NegH_p^R = \sum_{p=-\infty}^{p=+\infty} |\phi(p)|^2 Ln\left[|\phi(p)|^2\right] = -Ln\left(\frac{4\pi\hbar e^{2(1-\gamma)}}{L}\right) = -\lim_{n\to+\infty} H_p(n)$$

$$\overline{H}_p^R = -\sum_{p=-\infty}^{p=+\infty} \left[1 - |\phi(p)|^2\right] Ln\left[1 - |\phi(p)|^2\right]$$

$$H_p^M = -\sum_{p=-\infty}^{p=+\infty} i\left[1 - |\phi(p)|^2\right] Ln\left\{i\left[1 - |\phi(p)|^2\right]\right\}$$

$$H_p^C = -\sum_{p=-\infty}^{p=+\infty} Pc(p) Ln[Pc(p)] = -\sum_{p=-\infty}^{p=+\infty} 1 \times Ln[1] = -\sum_{p=-\infty}^{p=+\infty} (1 \times 0) = 0 = H_p^R + NegH_p^R$$

That means also and most importantly, for the wavefunction momentum distribution and in the probability set and universe $\mathcal{C} = \mathcal{R} + \mathcal{M}$, we have complete order, no chaos, no ignorance, no uncertainty, no disorder, no randomness, no information loss or gain but a conservation of information, and no unpredictability since all measurements are completely and perfectly deterministic ($Pc(p) = 1$ and $H_p^C = 0$).

The quantum mechanical entropic uncertainty principle states that for $x_0 p_0 = \hbar$ then:

$H_x^R + H_p^R(n) \geq Ln(e\pi) \cong 2.144729886$ ... nats, (base e in Ln gives the "natural units" nat).

For $x_0 p_0 = \hbar$, the sum of the position and momentum entropies yields:

$H_x^R + H_p^R(\infty) = Ln(8\pi e^{1-2\gamma}) \cong 3.069740098$ ... nats, (base e in Ln gives the "natural units" nat).

which satisfies the quantum entropic uncertainty principle.

The following figures (**Figures 38–51**) illustrate all the computations done above.

4. Conclusion and perspectives

In the current research work, the original extended model of eight axioms (*EKA*) of A. N. Kolmogorov was connected and applied to the infinite potential well problem in quantum mechanics theory. Thus, a tight link between quantum mechanics and the novel paradigm (*CPP*) was achieved. Consequently, the model of "Complex Probability" was more developed beyond the scope of my 19 previous research works on this topic.

Additionally, as it was proved and verified in the novel model, before the beginning of the random phenomenon simulation and at its end we have the chaotic factor (*Chf* and *MChf*) is zero and the degree of our knowledge (*DOK*) is one since the stochastic fluctuations and effects have either not started yet or they have terminated

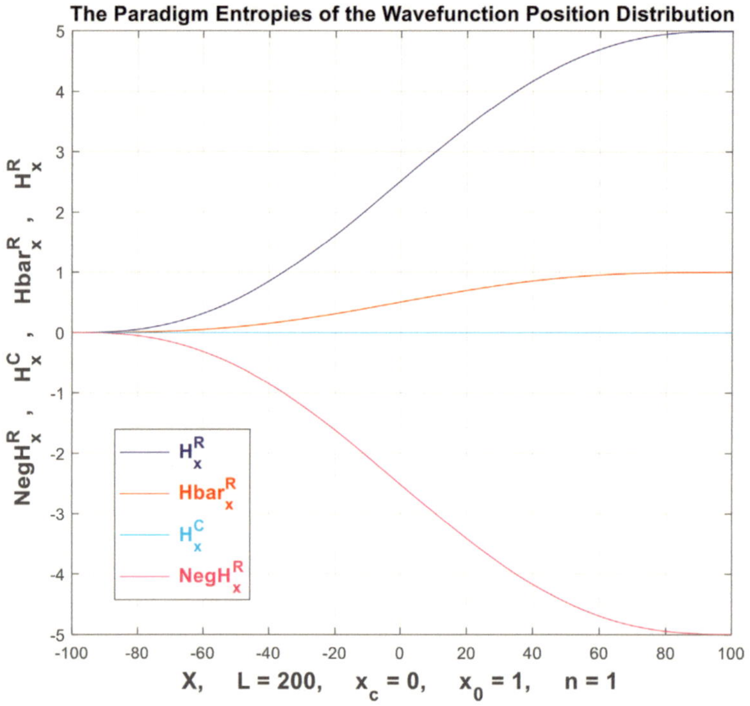

Figure 38.
The graphs of H_x^R,$\overline{H}_x^R$,H_x^C,$NegH_x^R$ *as functions of* X *for* $n = 1$.

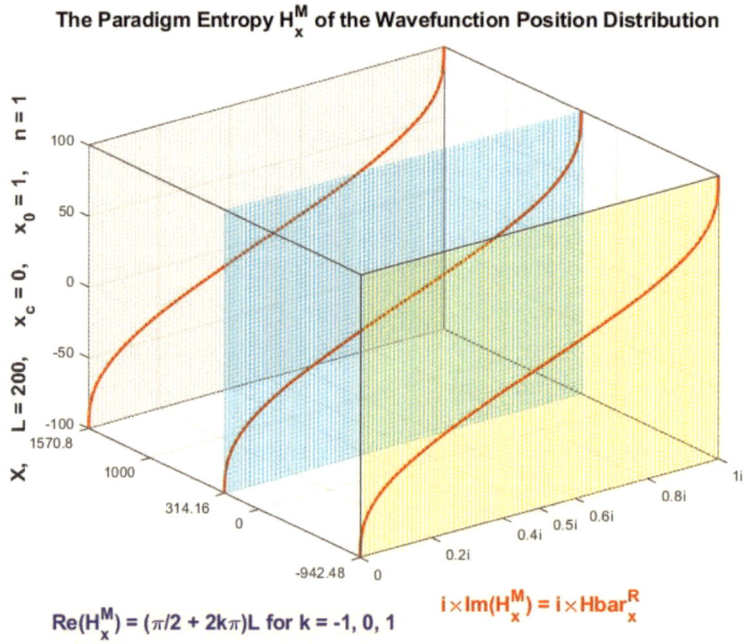

Figure 39.
The graph of $H_x^M = \text{Re}\left(H_x^M\right) + i\text{Im}\left(H_x^M\right)$ *in red as functions of* X *for* $n = 1$ *and for* $k = -1,0,1$ *in the planes in yellow, in cyan, and in light gray, respectively.*

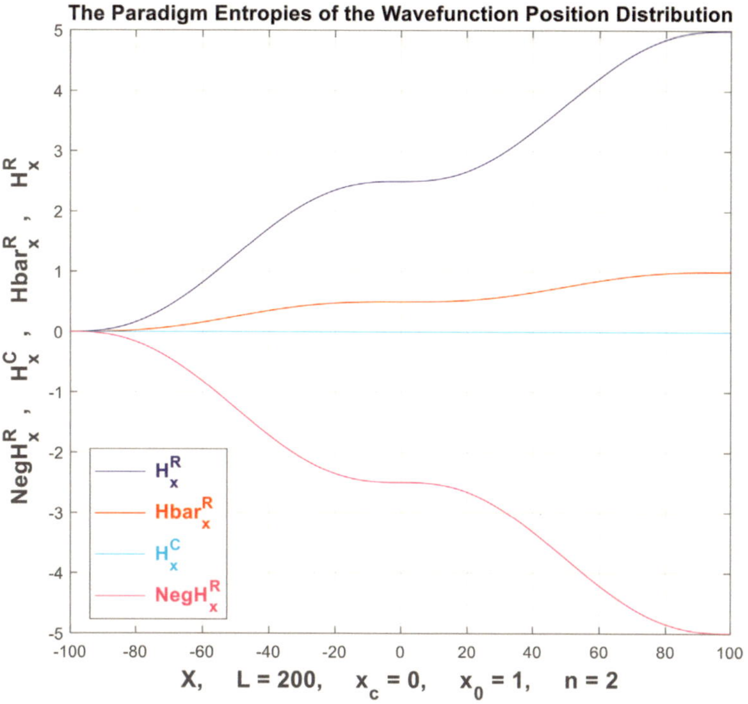

Figure 40.
The graphs of $H_x^R, \overline{H}_x^R, H_x^C, NegH_x^R$ *as functions of* X *for* $n = 2$.

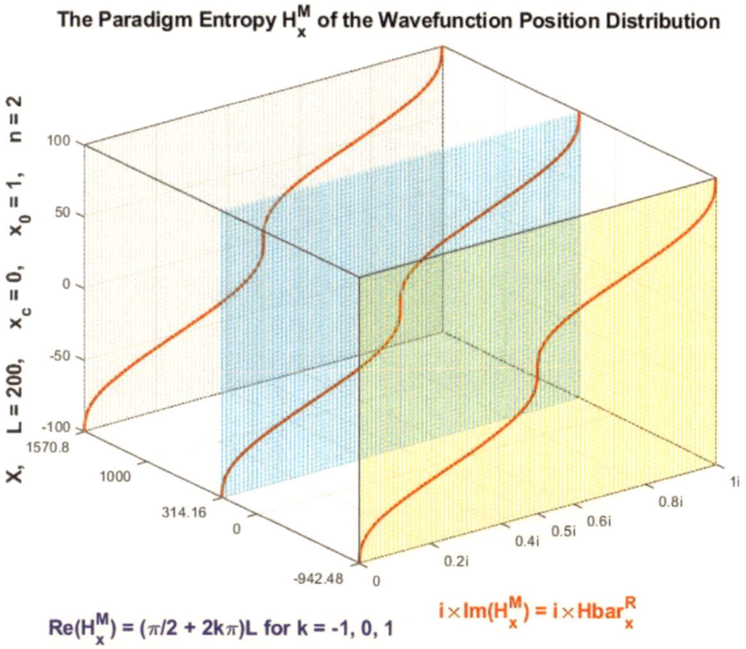

Figure 41.
The graph of $H_x^M = \text{Re}\left(H_x^M\right) + i\text{Im}\left(H_x^M\right)$ *in red as functions of* X *for* $n = 2$ *and for* $k = -1{,}0{,}1$ *in the planes in yellow, in cyan, and in light gray, respectively.*

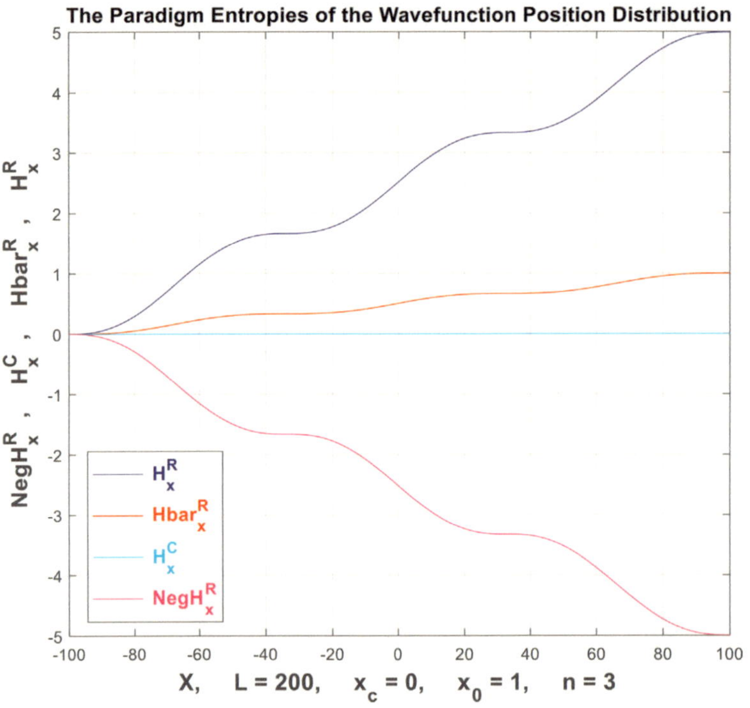

Figure 42.
The graphs of H_x^R,$\overline{H}_x^R$,H_x^C,$NegH_x^R$ as functions of X for $n = 3$.

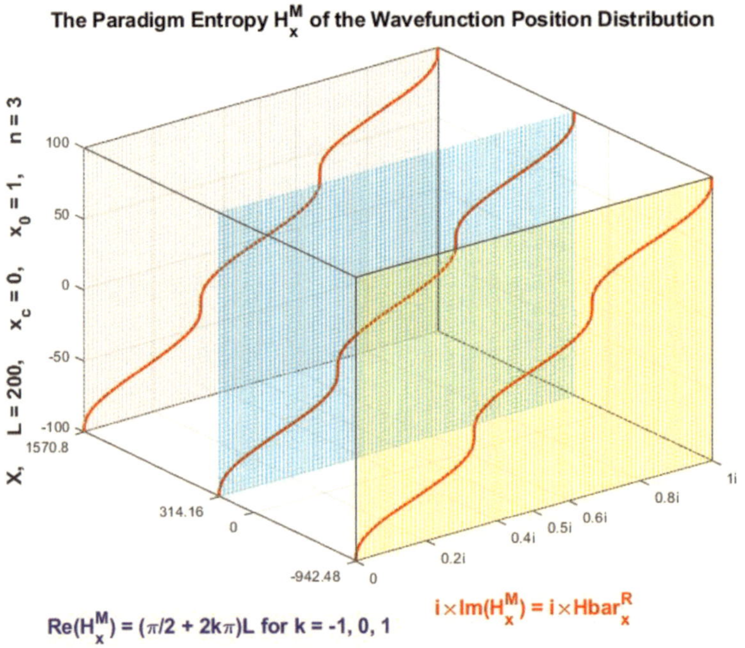

Figure 43.
The graph of $H_x^M = \text{Re}\left(H_x^M\right) + i\text{Im}\left(H_x^M\right)$ in red as functions of X for $n = 3$ and for $k = -1,0,1$ in the planes in yellow, in cyan, and in light gray, respectively.

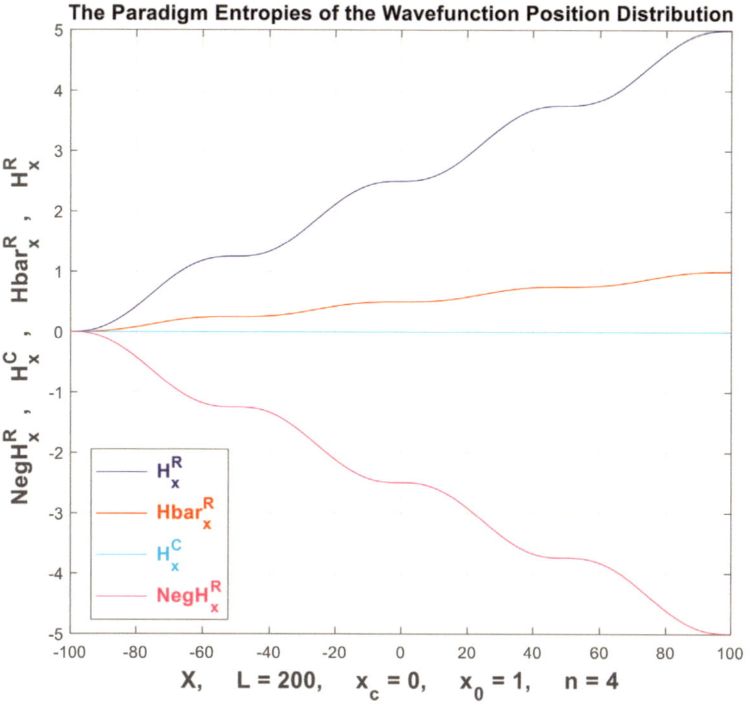

Figure 44.
The graphs of $H_x^R, \overline{H}_x^R, H_x^C, NegH_x^R$ *as functions of X for* $n = 4$.

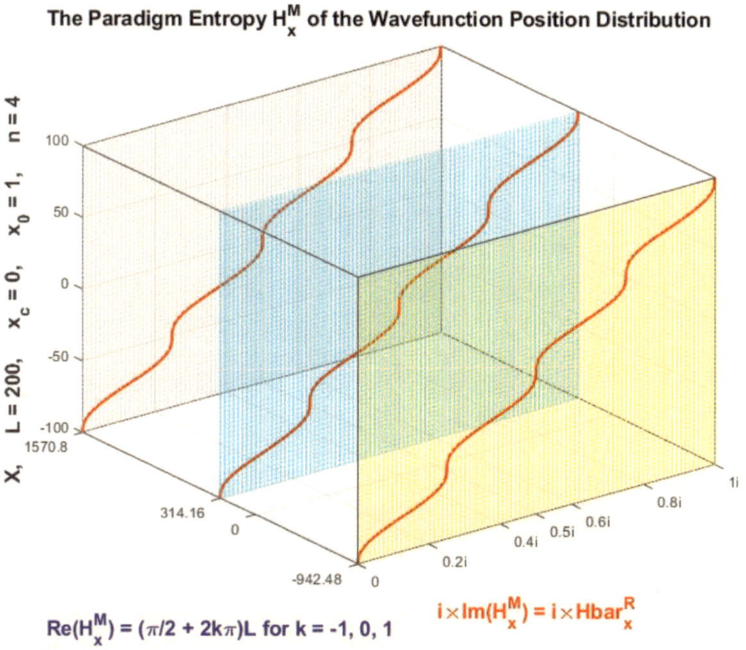

Figure 45.
The graph of $H_x^M = \text{Re}\left(H_x^M\right) + i\text{Im}\left(H_x^M\right)$ *in red as functions of X for* $n = 4$ *and for* $k = -1, 0, 1$ *in the planes in yellow, in cyan, and in light gray, respectively.*

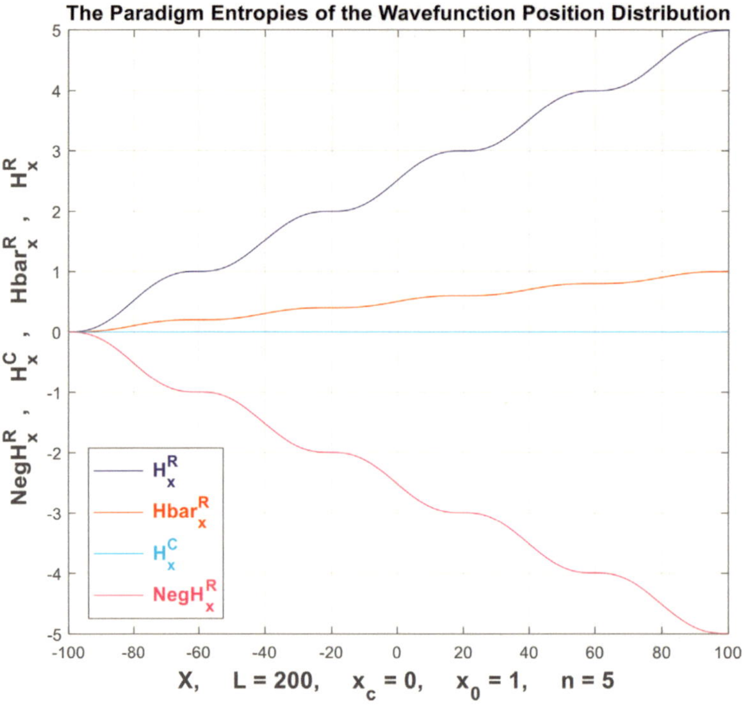

Figure 46.
The graphs of $H_x^R, \overline{H}_x^R, H_x^C, NegH_x^R$ *as functions of* X *for* $n = 5$.

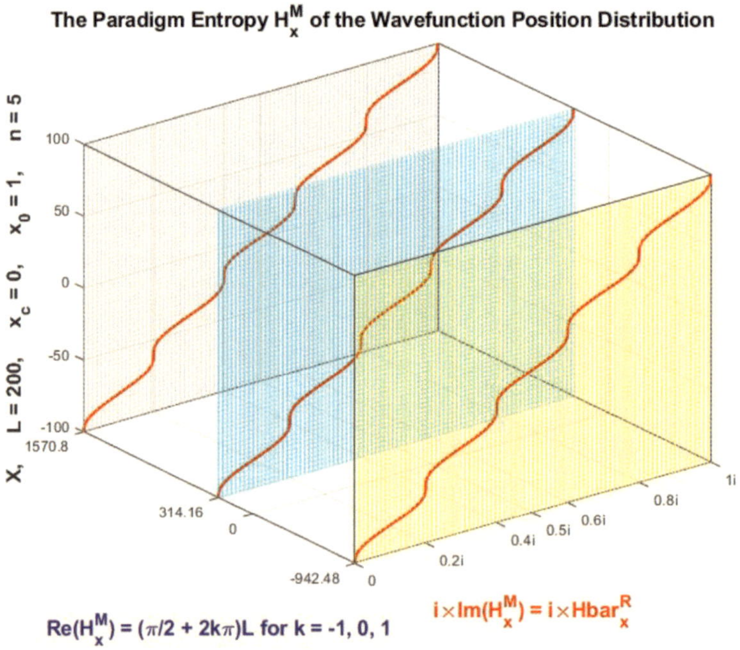

Figure 47.
The graph of $H_x^M = \mathrm{Re}\left(H_x^M\right) + i\mathrm{Im}\left(H_x^M\right)$ *in red as functions of* X *for* $n = 5$ *and for* $k = -1,0,1$ *in the planes in yellow, in cyan, and in light gray, respectively.*

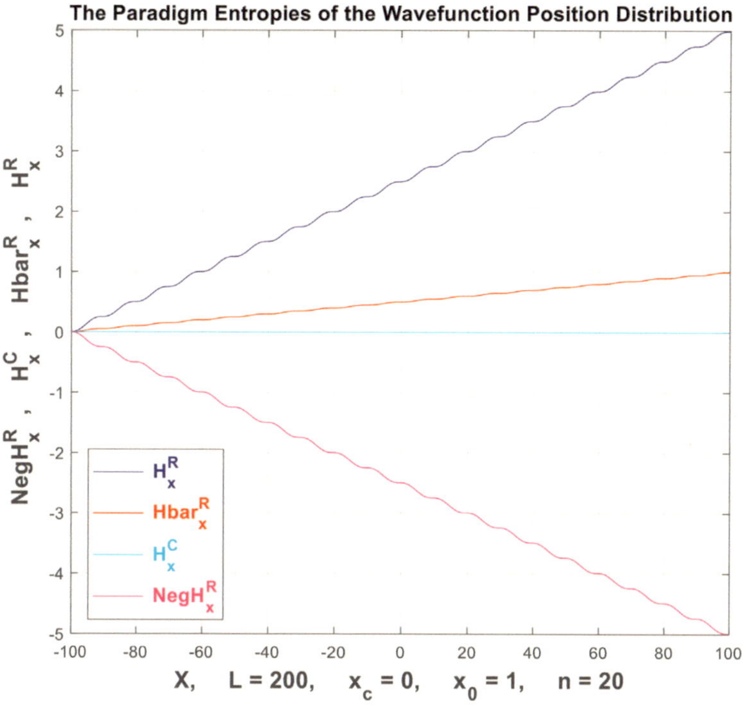

Figure 48.
The graphs of H_x^R,$\overline{H}_x^R$,H_x^C,$NegH_x^R$ *as functions of* X *for* $n = 20$.

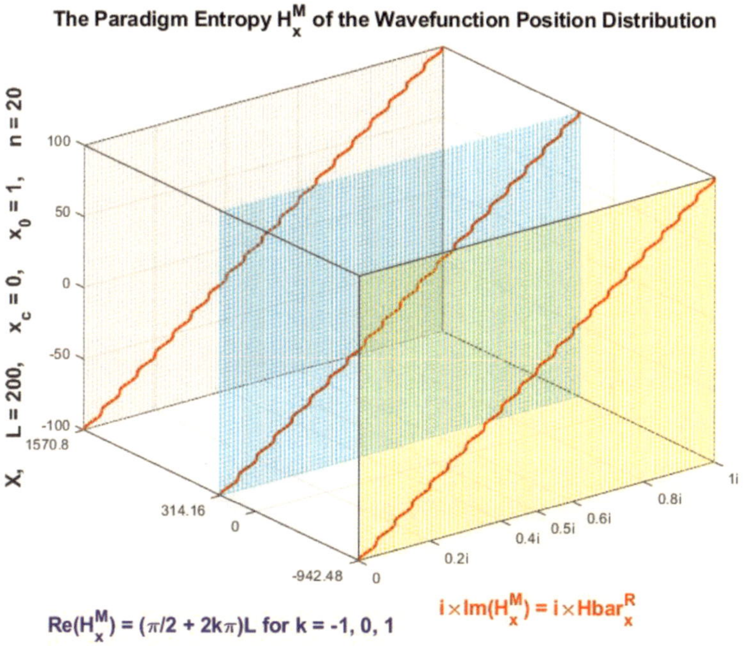

Figure 49.
The graph of $H_x^M = \text{Re}\left(H_x^M\right) + i\text{Im}\left(H_x^M\right)$ *in red as functions of* X *for* $n = 20$ *and for* $k = -1,0,1$ *in the planes in yellow, in cyan, and in light gray, respectively.*

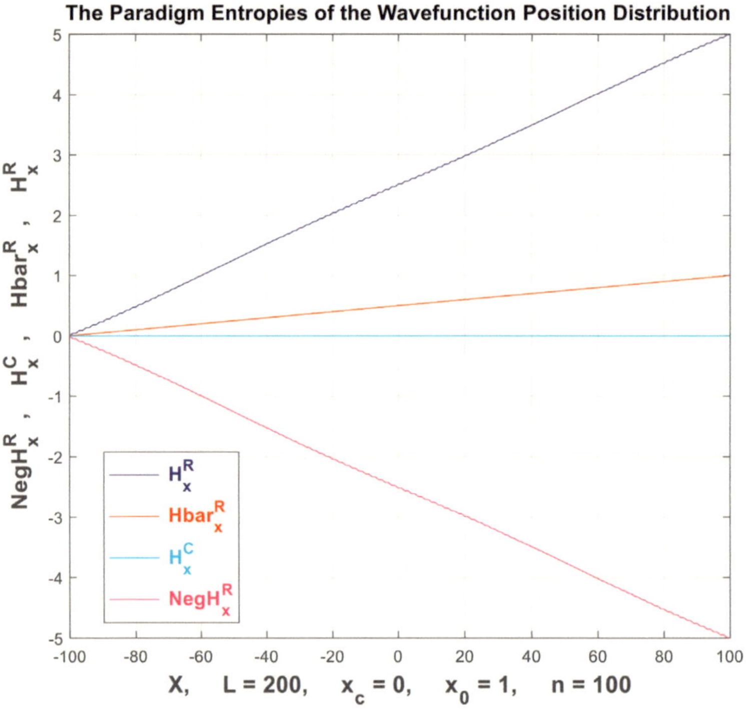

Figure 50.
The graphs of $H_x^R, \overline{H}_x^R, H_x^C, NegH_x^R$ *as functions of X for* $n = 100$.

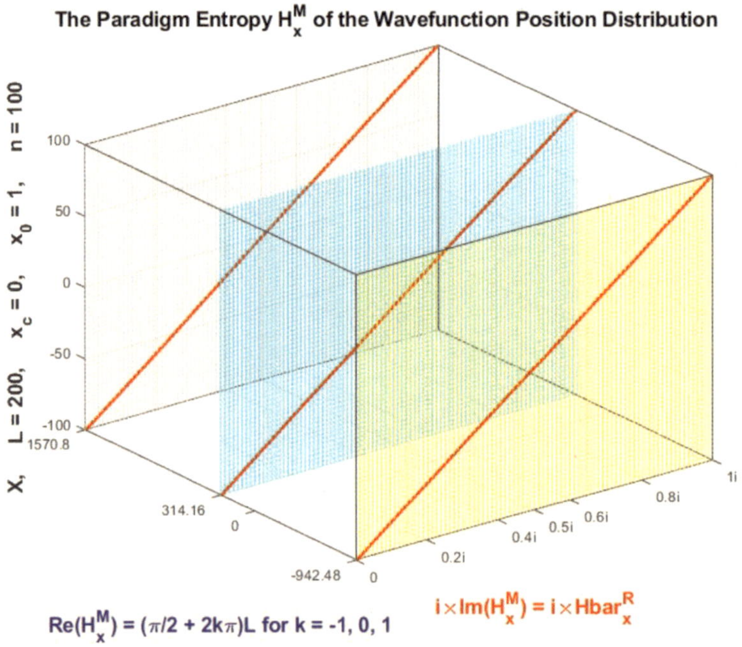

Figure 51.
The graph of $H_x^M = \text{Re}\left(H_x^M\right) + i\text{Im}\left(H_x^M\right)$ *in red as functions of X for* $n = 100$ *and for* $k = -1,0,1$ *in the planes in yellow, in cyan, and in light gray, respectively.*

and finished their task on the probabilistic phenomenon. During the execution of the nondeterministic phenomenon and experiment we also have: $0.5 \leq DOK < 1$, $-0.5 \leq Chf < 0$, and $0 < MChf \leq 0.5$. We can see that during this entire process we have incessantly and continually $Pc^2 = DOK - Chf = DOK + MChf = 1 = Pc$, that means that the simulation which behaved randomly and stochastically in the real set and universe $\mathcal{R}$ is now certain and deterministic in the complex probability set and universe $\mathcal{C} = \mathcal{R} + \mathcal{M}$, and this after adding to the random experiment executed in the real universe $\mathcal{R}$ the contributions of the imaginary set and universe $\mathcal{M}$ and hence after eliminating and subtracting the chaotic factor from the degree of our knowledge. Furthermore, the real, imaginary, complex, and deterministic probabilities and that correspond to each value of the momentum random variable P have been determined in the three probabilities sets and universes which are $\mathcal{R}$, $\mathcal{M}$, and $\mathcal{C}$ by P_r, P_m, Z and Pc respectively. Consequently, at each value of P, the novel quantum mechanics and *CPP* parameters P_r, P_m, P_m/i, *DOK*, *Chf*, *MChf*, *Pc*, and Z are surely and perfectly predicted in the complex probabilities set and universe $\mathcal{C}$ with Pc maintained equal to one permanently and repeatedly.

In addition, referring to all these obtained graphs and executed simulations throughout the whole research work, we are able to quantify and visualize both the system chaos and stochastic effects and influences (expressed and materialized by *Chf* and *MChf*) and the certain knowledge (expressed and materialized by *DOK* and *Pc*) of the new paradigm. This is without any doubt very fruitful, wonderful, and fascinating and proves and reveals once again the advantages of extending A. N. Kolmogorov's five axioms of probability and hence the novelty and benefits of my inventive and original model in the fields of prognostics, applied mathematics, and quantum mechanics that can be called verily: "The Complex Probability Paradigm".

As prospective research, we aim to develop the novel prognostic paradigm conceived and implement it in a large set of nondeterministic phenomena in quantum mechanics.

Author details

Abdo Abou Jaoudé
Department of Mathematics and Statistics, Faculty of Natural and Applied Sciences, Notre Dame University-Louaize, Lebanon

*Address all correspondence to: abdoaj@idm.net.lb

References

[1] Wikipedia, the free encyclopedia, Quantum Mechanics. Available from: https://en.wikipedia.org/

[2] Wikipedia, the free encyclopedia, Particle in a Box. Available from: https://en.wikipedia.org/

[3] Abou Jaoude A, El-Tawil K, Kadry S. Prediction in complex dimension using Kolmogorov's set of axioms. Journal of Mathematics and Statistics, Science Publications. 2010;**6**(2):116-124

[4] Abou Jaoude A. The complex statistics paradigm and the law of large numbers. Journal of Mathematics and Statistics, Science Publications. 2013;**9**(4):289-304

[5] Abou Jaoude A. The theory of complex probability and the first order reliability method. Journal of Mathematics and Statistics, Science Publications. 2013;**9**(4):310-324

[6] Abou Jaoude A. Complex probability theory and prognostic. Journal of Mathematics and Statistics, Science Publications. 2014;**10**(1):1-24

[7] Abou Jaoude A. The complex probability paradigm and analytic linear prognostic for vehicle suspension systems. American Journal of Engineering and Applied Sciences, Science Publications. 2015;**8**(1):147-175

[8] Abou Jaoude A. The paradigm of complex probability and the Brownian motion. Systems Science and Control Engineering, Taylor and Francis Publishers. 2015;**3**(1):478-503

[9] Abou Jaoude A. The paradigm of complex probability and Chebyshev's inequality. Systems Science and Control Engineering, Taylor and Francis Publishers. 2016;**4**(1):99-137

[10] Abou Jaoude A. The paradigm of complex probability and analytic nonlinear prognostic for vehicle suspension systems. Systems Science and Control Engineering, Taylor and Francis Publishers. 2016;**4**(1):99-137

[11] Abou Jaoude A. The paradigm of complex probability and analytic linear prognostic for unburied petrochemical pipelines. Systems Science and Control Engineering, Taylor and Francis Publishers. 2017;**5**(1):178-214

[12] Abou Jaoude A. The paradigm of complex probability and Claude Shannon's information theory. Systems Science and Control Engineering, Taylor and Francis Publishers. 2017;**5**(1):380-425

[13] Abou Jaoude A. The paradigm of complex probability and analytic nonlinear prognostic for unburied petrochemical pipelines. Systems Science and Control Engineering, Taylor and Francis Publishers. 2017;**5**(1):495-534

[14] Abou Jaoude A. The paradigm of complex probability and Ludwig Boltzmann's entropy. Systems Science and Control Engineering, Taylor and Francis Publishers. 2018;**6**(1):108-149

[15] Abou Jaoude A. The paradigm of complex probability and Monte Carlo methods. Systems Science and Control Engineering, Taylor and Francis Publishers. 2019;**7**(1):407-451

[16] Abou Jaoude A. Analytic prognostic in the linear damage case applied to buried petrochemical pipelines and the complex probability paradigm. Fault

Detection, Diagnosis and Prognosis, IntechOpen. 2020;**1**(5):65-103. DOI: 10.5772/intechopen.90157

[17] Abou Jaoude A. The Monte Carlo techniques and the complex probability paradigm. Forecasting in Mathematics - Recent Advances, New Perspectives and Applications, IntechOpen. 2020;**1**(1): 1-29. DOI: 10.5772/intechopen.93048

[18] Abou Jaoude A. The Paradigm of Complex Probability and Prognostic Using FORM. London Journal of Research in Science: Natural and Formal (LJRS), London Journals Press. 2020; **20**(4):1-65. DOI: 10.17472/LJRS

[19] Abou Jaoude A. The paradigm of complex probability and the central limit theorem. London Journal of Research in Science: Natural and Formal (LJRS), London Journals Press. 2020;**20**(5):1-57. DOI: 10.17472/LJRS

[20] Abou Jaoude A. The Paradigm of Complex Probability and Thomas Bayes' Theorem. The Monte Carlo Methods - Recent Advances, New Perspectives and Applications. IntechOpen. 2021. DOI: 10.5772/intechopen.98340

[21] Abou Jaoude A. The paradigm of complex probability and Isaac Newton's classical mechanics: On the foundation of statistical physics. The Monte Carlo Methods - Recent Advances, New Perspectives and Applications. IntechOpen. 2021. DOI: 10.5772/ intechopen.98341

[22] Abou Jaoude A. The Computer Simulation of Monté Carlo Methods and Random Phenomena. United Kingdom: Cambridge Scholars Publishing; 2019

[23] Abou Jaoude A. The Analysis of Selected Algorithms for the Stochastic Paradigm. United Kingdom: Cambridge Scholars Publishing; 2019

[24] Abou Jaoude A. The Analysis of Selected Algorithms for the Statistical Paradigm. Vol. 1. The Republic of Moldova: Generis Publishing; 2021

[25] Abou Jaoude A. The Analysis of Selected Algorithms for the Statistical Paradigm. Vol. 2. The Republic of Moldova: Generis Publishing; 2021

[26] Abou Jaoude A. Forecasting in Mathematics - Recent Advances, New Perspectives and Applications. London. United Kingdom: IntechOpen; 2021

[27] Abou Jaoude A. The Monte Carlo Methods - Recent Advances, New Perspectives and Applications. London. United Kingdom: IntechOpen; 2022

[28] Abou Jaoude A. Ph.D. Thesis in Applied Mathematics: Numerical Methods and Algorithms for Applied Mathematicians. Spain: Bircham International University; 2004. Available from: http:// www.bircham.edu

[29] Abou Jaoude A. Ph.D. Thesis in Computer Science: Computer Simulation of Monté Carlo Methods and Random Phenomena. Spain: Bircham International University; 2005. Available from: http:// www.bircham.edu

[30] Abou Jaoude A. Ph.D. Thesis in Applied Statistics and Probability: Analysis and Algorithms for the Statistical and Stochastic Paradigm. Spain: Bircham International University; 2007. Available from: http://www.bircha m.edu

Chapter 3

Stability of Algorithms in Statistical Modeling

Alexander A. Kronberg and Tatiana K. Kronberg

Abstract

In this paper, we investigate algorithms stability for calculation of multidimensional integrals using the statistical modeling methods. We considered issues of the algorithms optimization and we give sufficient conditions for the stability. We apply our approach to both calculation of integral from the regression function and the moments integral calculation. In all our numerical experiences, we used the mt19937 pseudorandom number generator.

Keywords: statistical modeling, pseudorandom numbers, optimal density, integral estimation, Monte Carlo methods

1. Introduction

One of the main problem of the statistical modeling method (the Monte Carlo method) is the problem of quality for pseudorandom numbers. In the paper, we consider a task of multidimensional integrals calculation by the statistical modeling method and give sufficient conditions for the stability of this task to quality of pseudorandom numbers. Included results of various numerical experiences with the mt19937 pseudorandom number generator. In our work, we discuss important issues of algorithms optimization in the statistical modeling. In particular, we apply the new approach to the following: a task of finding of integral functionals from solution of boundary-value problems for both the linear [1] or nonlinear [2] elliptic equations (the estimations are given near to a boundary).

The paper is organized as follows: In Section 2, we give the sufficient conditions of stability. Calculation of an integral of very large dimensions is discussed in Section 3. Rare events effect is the subject of Section 4. In Section 5, we describe calculation of integral moments. In Section 6, we apply our approach to calculate an integral of the regression function. In Section 7, we give the conclusion of our studies.

2. Sufficient conditions of stability

Let

$$I = \int_D f(x)dx \tag{1}$$

be the Riemann integral. Here D is a domain of the s-dimensional Euclidean space R^s. If the dimension s is large enough then we must use a statistical modeling method. In this case, our integral has form of the mathematical expectation for a random value $\eta = f(\xi)/p(\xi)$:

$$\int_D f(x)dx = \int_D p(x)\frac{f(x)}{p(x)}dx = \mathcal{E}\frac{f(x)}{p(x)} = \mathcal{E}\eta. \tag{2}$$

Here $p(x)$ is a density of random variable $\xi \in D$. We put $p(x) \neq 0$ for $f(x) \neq 0$, and we say that there exists integral

$$\int_D |f(x)|dx.$$

A variance of the random value η:

$$\sigma^2 = \mathbf{var}\eta = \mathcal{E}\eta^2 - (\mathcal{E}\eta)^2 = \int_D p(x)\left[\frac{f(x)}{p(x)}\right]^2 dx - I^2 = \int_D \frac{f^2(x)}{p(x)}dx - I^2. \tag{3}$$

We estimate the mathematical expectation $\mathcal{E}\eta$ by the sum $\sum_{i=1}^N \eta_i/N$, where η_i are independent realizations of the random value η. Suppose σ^2 is finite, and N is large enough; then from the classical central limit theorem, it follows that the random value $\sum_{i=1}^N \eta_i/N$ has distribution close to the normal distribution with a mathematical expectation I, and mean-square deviation $\sigma/\sqrt{N}$. This property above is useful to estimate error, e.g., using the 3σ rule. So we have

$$\left| I - \frac{1}{N}\sum_{i=1}^N \eta_i \right| \leq \frac{3\sigma}{\sqrt{N}} \tag{4}$$

with probability 0,997, approximately.

Let us η_i be realizable sampling values; then the value σ is estimated as the following:

$$\sigma^2 \approx \frac{1}{N}\sum_{i=1}^N \eta_i^2 - \left(\frac{1}{N}\sum_{i=1}^N \eta_i\right)^2. \tag{5}$$

Suppose $\mathcal{E}\eta^4$ is finite and in Eq. (3) we replace the σ by its approximate value. Then it changes the estimation of error in calculation of the integral in order of $O\left(\frac{1}{N}\right)$.

In practice, when we simulate random variables ξ, we receive simulation with some density $q(x)$ instead of simulation with the origin density $p(x)$. Now, we investigate the stability of the theoretical estimation Eq. (2). Let us consider the following expression:

$$\int_D p(x)\frac{f(x)}{p(x)}dx - \int_D p(x)\frac{q(x)}{p(x)}dx = \int_D \frac{f(x)}{p(x)}(p(x) - q(x))dx \leq$$
$$\leq \varepsilon \int_D \frac{|f(x)|}{p(x)}dx, \tag{6}$$

where $\varepsilon = \sup_{x \in D} |p(x) - q(x)|$. By $I(|f|/p)$ denote the integral $\int \frac{|f(x)|}{p(x)} dx$. The both values ε and $I(|f|/p)$ provide the guaranteed proximity of the real estimation to the theoretical one of the integral.

Example 1. The inequality Eq. (6) is reduced to the equality if $D = D_1 \cup D_2, D_1 \cap D_2 = \varnothing$. For $x \in D_1$ we get $p(x) - q(x) \equiv \varepsilon > 0$, and $f(x) \equiv 0$ in D_2. If the condition $I(|f|/p) = +\infty$ holds, then the error of the real estimation of the integral will be infinity for any $\varepsilon > 0$.

Hence, a quality of pseudorandom variables (i.e., smallness of ε) does not yet guaranties the smallness of the error in general, as the integral $I(|f|/p)$ have to be both finite and not very great in magnitude.

Suppose we simultaneously make the estimations for both $I(|f|/p)$ and the origin integral Eq. (2) using the same density $p(x)$. Then we need to ask boundedness of the integral $\int_D \frac{|f(x)|}{p^2(x)} dx$ to get the guaranteed stability of the estimation for the integral $I(|f|/p)$, and so on. The qualitative comparison of simulation with both densities $p_1(x)$ and $p_2(x)$ can be provided not only by a magnitude of the variance estimation (here we do not pay attention to the complexity of random values simulation) but also magnitudes of both the integrals $I(|f|/p_1)$ and $I(|f|/p_2)$. On the other hand we have the Schwarz inequality:

$$\int_D \frac{f(x)}{p(x)} (p(x) - q(x)) dx \le \sqrt{\int_D [p(x) - q(x)]^2 dx} \sqrt{\int_D \frac{f^2(x)}{p^2(x)} dx}. \tag{7}$$

The sufficient condition for the estimation to be stability is that $\int_D \frac{f^2(x)}{p^2(x)}$ to be finite and not very great. The Schwarz inequality is reduced to the equality if and only if

$$\lambda \frac{f(x)}{p(x)} = p(x) - q(x), \tag{8}$$

where λ is a real number. From the above we get

$$\begin{gathered} q(x) = p(x) - \lambda \frac{f(x)}{p(x)}, \\ \int_D q(x) dx = \int_D p(x) dx - \lambda \int_D \frac{f(x)}{p(x)} dx. \end{gathered} \tag{9}$$

Therefore, the equality in Eq. (7) is reached under the necessary condition $\int_D \frac{f(x)}{p(x)} dx = 0$, when $\int_D p(x) dx = \int_D q(x) dx = 1$.

Example 2. The condition above is realized, e.g., if

$$\begin{gathered} D = [-1, 1], \quad D_1 = [-1, 0], \quad D_1 = [0, 1], \\ f(x) = -1 \, \text{in} \, D_1, \quad f(x) = 1 \, \text{in} \, D_2, \quad p(x) = p(-x). \end{gathered}$$

Let us η_k be $[f(\xi)/p(\xi)]^k$. Now we consider an estimation:

$$\mathcal{E}\eta_k = \int_D p(x) \left[\frac{f(x)}{p(x)}\right]^k dx = \int_D \frac{f^k(x)}{p^{k-1}(x)} dx, \quad k \ge 1. \tag{10}$$

This expectation is actually estimated by the integral: $\int_D q(x)[f(x)/p(x)]^k dx$:

$$\int_D \frac{f^k(x)}{p^{k-1}(x)}dx - \int_D q(x)\frac{f^k(x)}{p^k(x)}dx = \int_D [p(x)-q(x)]\frac{f^k(x)}{p^k(x)}dx \leq \\ \leq \varepsilon \int_D \frac{|f^k(x)|}{p^k(x)}dx. \tag{11}$$

The last integral is assumed to be a finite, and not very large. These conditions are desirable. In Eq. (11) the equality is reached like to the Example 1.

For all cases above, the stability will be observed if $|f(x)/p(x)| \leq M < +\infty$ for not very great M. From the Schwarz inequality we have:

$$\int_D [p(x)-q(x)]\frac{f^k(x)}{p^k(x)}dx = \int_D \frac{p(x)-q(x)}{p^\beta(x)} \cdot \frac{f^k(x)}{p^{k-\beta}(x)}dx \leq \\ \leq \sqrt{\int_D \frac{[p(x)-q(x)]^2}{p^{2\beta}(x)}dx}\sqrt{\int_D \frac{f^{2k}(x)}{p^{2(k-\beta)}(x)}dx}, \tag{12}$$

where β is a real number. We get a family of proximity measures for the distribution densities:

$$\int_D \frac{[p(x)-q(x)]^2}{p^{2\beta}(x)}dx. \tag{13}$$

For $\beta = 0,5$ we obtain expression

$$\chi^2(p,q) = \int_D \frac{[p(x)-q(x)]^2}{p^{(x)}dx.} \tag{14}$$

that is well known in the mathematical statistics.

For $\beta = -0,5$ we get $\int_D [p(x)-q(x)]^2 p(x)dx$ and have the obvious inequalities

$$\int_D [p(x)-q(x)]^2 p(x)dx \leq \sup_{x\in D} p(x) \cdot \int_D [p(x)-q(x)]^2 dx,$$

$$\int_D [p(x)-q(x)]^2 p(x)dx \leq \sup_{x\in D}[p(x)-q(x)]^2 \int_D p(x)dx = \sup_{x\in D}[p(x)-q(x)]^2.$$

In Eq. (6) the equality is satisfied if and only if

$$\frac{p-q}{p^\beta} = \lambda\frac{f^k}{p^{k-\beta}}, \quad p-q = \lambda\frac{f^k p^\beta}{p^{k-\beta}}, \tag{15}$$

i.e., the necessary condition is $\int_D \frac{f^k(x)}{p^{k-2\beta}(x)}dx = 0$. This is realized in the Example 2.

Let us remark that for $k = 1$ and $\beta = 1$ we have

$$\int_D \frac{p(x)-q(x)}{p(x)}f(x)dx = \int_D \frac{p-q}{\sqrt{p}} \cdot \frac{f}{\sqrt{p}} \leq \\ \leq \sqrt{\int_D \frac{(p-q)^2}{p}dx}\sqrt{\int_D \frac{f^2}{p}dx} = \sqrt{\chi^2(p,q)} \cdot \mathcal{E}\eta_2. \tag{16}$$

We assume that integrals $\int_D [f(x) + \varepsilon_i(x)]dx$ are known and the subintegral functions $f(x) + \varepsilon_i(x) > 0$ close to a function $f(x) \geq 0$. Suppose also

$$1 - \delta_1(\varepsilon_i) < \frac{f(x)}{f(x) + \varepsilon_i(x)} < 1 + \delta_2(\varepsilon_i),$$
$$J(f) - \delta_3(\varepsilon_i) \leq J(f + \varepsilon_i) \leq J(f) + \delta_4(\varepsilon_i), \quad \delta_j \to 0, \text{ as } \varepsilon_i \to 0, \ j = 1, 2, 3, 4. \quad (17)$$
$$\int_D q(x)\frac{f(x)}{p(x)}dx = \int_D q(x)\frac{f(x)J(f + \varepsilon_i)}{f(x) + \varepsilon_i(x)}dx = \mathcal{E}\hat{\eta},$$

where the random value $\hat{\eta}$ has a form:

$$\hat{\eta} = \frac{f(\hat{\xi})J(f + \varepsilon_i)}{f(\hat{\xi}) + \varepsilon_i(\hat{\xi})}. \quad (18)$$

Here $\hat{\xi}$ is distributed with the density $q(x)$. Keeping the above factors in mind we get the following:

$$[J(f) - \delta_3][1 - \delta_1] \leq \eta \leq [1 + \delta_2][J(f) + \delta_4];$$

i. Regardless of $q(x)$, i.e., regardless of quality of a pseudorandom number generator we have $\mathcal{E}\hat{\eta} \to J(f), \boldsymbol{var}\hat{\eta} \to 0, \text{ as} \varepsilon \to 0;$

ii. All moments $\mathcal{E}\hat{\eta}^k$ of the random variable are finite.

Let we calculate $\int_D \frac{|f(x)|}{p(x)} dx$ using the density

$$p_1(x) = \frac{|f(x)|}{I(|f|/p)p(x)},$$

then the estimation variance equals to zero.

Suppose we calculate the integral $\int_D f(x)dx$ with the density $p_1(x)$. In this case it would be interesting to know both the values $\int_D \frac{|f(x)|}{p_1} dx$ and $\int_D \frac{f^2}{p_1^2} dx$.

Proposition 1. $I(|f|/p_1) = I(|f|/p)$.

Proposition 2. $I(f^2/p_1^2) = I^2(|f|/p)I(p^2)$.

Now we consider the density

$$p_2(x) = \frac{f^2(x)}{p(x)I(f^2/p)}.$$

Using the density above for the estimation of the integral $\int_D \frac{f^2}{p} dx$ we obtain the estimation variance equals to zero.

Further we estimate the integral $\int_D f(x)dx$ with the density p_2: $\hat{\eta} = f(\xi_2)/p(\xi_2)$, ξ_2 is distributed with $p_2(x)$. Suppose $\eta = f(\xi)/p(\xi)$, ξ is distributed with $p(x)$; then $I(f^2/p_2) = I(f^2/p)$.

Proposition 3. $\boldsymbol{var}\,\eta = \boldsymbol{var}\,\hat{\eta}$.

Proposition 4. $\left[\int\limits_D \frac{f(x)}{p(x)}dx\right]^2 \leq (vol D)\cdot \int\limits_D \frac{f^2(x)}{p^2(x)}dx,$ where $vol D$ is the volume of the domain D.

Proposition 5. If $vol D = 1, p_3(x) = \frac{f^2(x)}{I(f^2)}, \ \eta_3 = \frac{f(\xi_3)}{p(\xi_3)},$ where ξ_3 is a random variable distributed with the density of $p_3(x)$; $\eta_4 = \frac{f(\xi_4)}{p(\xi_4)},$ where ξ_4 is a random variable distributed with the density of $p(x) \equiv 1,$ then $\boldsymbol{var}\eta_3 = \boldsymbol{var}\eta_4.$

In actual practice normalization constants are usually unknown for both $p_1(x)$ and $p_2(x)$. But using densities close to them we can get the approximate equalities in the Prepositions 1, 2, 3.

Now we consider

$$I(f) = \int\limits_{[0,1]^{10}} x_1 x_2 \dots x_{10}\, dx_1 \dots dx_{10}, \tag{19}$$

where the integration domain $D = [0,1]^{10}$ is the 10-dimensional unit cube, the subintegral function $f(x)$ is equals to $x_1x_2\dots x_{10}$. To realize algorithms of the statistical modeling at a computer it is necessary to set a number N of realizations for random variable $\eta = f(\xi)/p(\xi)$, where ξ is distributed with the density $p(\xi)$. In fact we realize the discrete set of numbers $\xi_i, \ i = 1, \dots, N,$ which we can consider to be realizations of some distribution $q_N(x)$.

In all our numerical computations we use the pseudorandom number generator: *generator type* mt19937 [3]. For $s = 10, \ p(x) \equiv 1$ we have

$$I(f) = (1/2)^{10} \approx 9{,}7656\cdot 10^{-4}, \quad I(f/p) = I(f),$$
$$I(f^2/p) = I(f^2/p^2) = (1/3)^{10} \approx 1{,}6935\cdot 10^{-5},$$
$$\sigma = \sqrt{(1/3)^{10} - (1/2)^{20}} \approx 3{,}998\cdot 10^{-3}.$$

Table 1 shows the empirical estimations $\hat{I}$ and $\hat{\sigma}$ for I and σ, respectively. Taking $p(x_i) = 3x_i^2$ over each coordinate we get

$$I(f^2/p) = (1/3)^{10}, \quad I(f/p) = \infty, \quad I(f^2/p^2) = \infty,$$
$$\sigma = \sqrt{(1/3)^{10} - (1/2)^{20}}.$$

The value σ is the same as one for $p(x) \equiv 1$.

N	$\hat{I}$	$\hat{\sigma}$
1,000,000	$9{,}785\cdot 10^{-4}$	$3{,}991\cdot 10^{-3}$
9,000,000	$9{,}768\cdot 10^{-4}$	$4{,}001\cdot 10^{-3}$
81,000,000	$9{,}766\cdot 10^{-4}$	$3{,}997\cdot 10^{-3}$
100,000,000	$9{,}765\cdot 10^{-4}$	$3{,}995\cdot 10^{-3}$

Table 1.
The results of numerical calculations for the integral Eq. (19) with the uniform density.

N	$\hat{I}$	$\hat{\sigma}$
1,000,000	$9,766 \cdot 10^{-4}$	$3,276 \cdot 10^{-3}$
9,000,000	$9,754 \cdot 10^{-4}$	$3,619 \cdot 10^{-3}$
81,000,000	$9,765 \cdot 10^{-4}$	$3,759 \cdot 10^{-3}$
100,000,000	Inf	nan

Table 2.
The results of numerical calculations for the integral Eq. (19) with the density $p(x_i) = 3x_i^2$.

For $N = 100000000$ the computer code outputs an error because of machine zero divide. The reason of this event is $\eta = \prod_1^{10}\left(1/\left(3\sqrt[3]{\alpha_i}\right)\right)$, where α_i are the pseudorandom numbers with the uniform density in $(0,1)$ [3]. If the formulas of random numbers simulation generate division by very small numbers then such formulas are one more source of the algorithms instability in the statistical modeling. As seen in **Table 2** the value of integral $I(f)$ is successfully estimated, but the empirical estimations of $\hat{\sigma}$ are sufficiently different from the theoretical value σ. This result is explained by the following: $\mathcal{E}\eta^4 = \infty$, $I(f/p) = \infty$, $I\left(f^2/p^2\right) = \infty$. Taking $p(x_i) = 2,6x_i^{1,6}$ we obtain $\sigma \approx 1,223 \cdot 10^{-3}$, $I(f/p) \approx 0,67556$, the finite value of $\mathcal{E}\eta^4$, and $I\left(f^2/p^2\right) = \infty$. Although the last estimation is infinite, but **Table 3** shows that both values $I(f)$ and $\hat{\sigma}$ are successfully calculated. The value of $\hat{\sigma}$ is very close to σ.

3. Integrals of very large dimensions

We are coming now to the question of calculation of an integral

$$I(f) = \int_0^\infty \dots \int_0^\infty e^{-(x_1+x_2+\dots+x_s)} dx_1 dx_2 \dots dx_{10} = 1 \quad (20)$$

with the distribution density $p(x) = \lambda^s e^{-\lambda(x_1+x_2+\dots+x_s)}$. For $\lambda \geq 2$, the estimation variance η is infinity. For $0 < \lambda < 2$, the variance will be finite. For $\lambda > 1$, we obtain $I(f/p) = \infty$ and $I\left(f^2/p^2\right) = \infty$. However, as seen in **Table 4**, the results of calculations for $N = 10000$ allow us to make the conclusion below. If we have the pseudorandom generator of the high quality and a good $p(x)$ then we can calculate the very high dimensional integrals.

N	$\hat{I}$	$\hat{\sigma}$
1,000,000	$9,769 \cdot 10^{-4}$	$1,212 \cdot 10^{-3}$
9,000,000	$9,760 \cdot 10^{-4}$	$1,220 \cdot 10^{-3}$
81,000,000	$9,766 \cdot 10^{-4}$	$1,223 \cdot 10^{-3}$

Table 3.
The results of numerical calculations for the integral Eq. (19) with the density $p(x_i) = 2,6x_i^{1,6}$.

λ	s	$\hat{I}$	$\hat{\sigma}$	σ
1,01	1000	0,999	0,320	0,324
1,01	10,000	0,995	1,39	1,31
1005	20,000	1004	0,807	0,805
1003	40,000	0,994	0,649	0,658
1001	80,000	0,991	0,605	0,661

Table 4.
The results of numerical calculations for the integral Eq. (20).

4. Special integrals

Let us consider the following class of the integrals:

$$\int\limits_{[0,+\infty)^s} f(x_1, x_2, \dots, x_s)\, dx_1 dx_2 \dots dx_s, \tag{21}$$

We define the behavior of the subintegral function as follows: $f(x_1, x_2, \dots, x_s)$ to be [label = ()]

1. close to 1 in the cube $[0, a]^s$ as $0 < a < 1$;

2. much less than unit as $a < x_i < b, \;\; b \geq a$;

3. equil to zero as $b \leq x_i < \infty$.

Note that very often the integration of functions can be reduced to the linear combination of the integrals similar to Eq. (21) using various replacements of variables.

Below, let us perform a theoretical and numerical analysis how to integrate a model function from our class. The model function is assumed to be $f \equiv 1$ as $0 \leq x_i \leq a$, otherwise $f = 0$. We take both distribution densities set $p_1(x_i) = \lambda e^{-\lambda x_i}, \;\; 0 \leq x_i < \infty$ and $p_2(x_i) = (\omega + 1)(1 - x_i)^{\omega}, \;\; 0 \leq x_i < 1$ to be examined. Our goal is to determine what of two densities provides the best accuracy of the integral computation with given model function.

If we simulate a random point $\xi = (\xi_1, \dots, \xi_s)$ with densities $p_1(x_i) = \lambda e^{-\lambda x_i}$ then the integral estimation is given by

$$\eta_s = \prod_{i=1}^{s} \lambda e^{\lambda \xi_i},$$
$$\mathbf{var}\,\eta_s = \mathcal{E}\eta_s^2 - (\mathcal{E}\eta_s)^2 = \left[\frac{1}{\lambda}\int\limits_0^a e^{\lambda x} dx\right]^s - a^s = \left[\frac{e^{\lambda a} - 1}{\lambda^2}\right]^s - a^s. \tag{22}$$

Testing the variance $\mathbf{var}\,\eta_s$ for the extremum over λ we get the minimum condition

$$\lambda a e^{\lambda a} - 2e^{\lambda a} + 2 = 0. \tag{23}$$

Let A be λa; then the equation Eq. (23) is reduced to

$$Ae^A - 2e^A + 2 = 0. \tag{24}$$

The equation above has the unique root at $A \approx 1,593620$. It follows that $\lambda_{min} = A/a$. For such λ_{min} the relative error with the 3σ rule is given by

$$\frac{3\sigma}{\sqrt{N}a^s} = 3\left(\frac{\left(e^A - 1\right)^s}{A^{2s}} - 1\right)^{0,5} / \sqrt{N}. \tag{25}$$

Suppose $N = 9 \cdot 10^6, \ \ s = 10, \lambda = A/a$; then the theoretical value of the relative error is approximately $8,72 \cdot 10^{-3}$. The numerical estimation of the relative error is approximately $8,69 \cdot 10^{-3}$ as $a \in [0,1; 0,001]$. Thus, the numerical estimation of one gives a good fit to the predicted value over a wide range of a.

Now we discuss the use of the density $p_2(x_i) = (\omega + 1)(1 - x_i)^\omega$. First, we estimate the second moment of a random value η_s:

$$\mathcal{E}\eta_s^2 = \left[\frac{1 - (1-a)^{1-\omega}}{1 - \omega^2}\right]^s. \tag{26}$$

The parameter ω is chosen to be A/a; then the expression above is rewritten as follows

$$\mathcal{E}\eta_s^2 = \left[\frac{1}{1 - A^2/a^2}\left(1 - (1-a)^{1-A/a}\right)\right]^s. \tag{27}$$

Let us consider $\mathcal{E}\eta_s^2$ as $a \to 0$:

$$\lim_{a\to 0} \mathcal{E}\eta_s^2 = \lim_{a\to 0}\left[\frac{a^2}{a^2 - A^2}\left(1 - (1-a)^{1-A/a}\right)\right]^s =$$
$$= -\lim_{a\to 0}\left[\frac{a^2}{A^2}\left(1 - (1-a)(1-a)^{-A/a}\right)\right]^s = -\lim_{a\to 0}\left[\frac{a^2}{A^2}\left(1 - (1-a)\left[(1-a)^{-1/a}\right]^A\right)\right]^s =$$
$$= -\lim_{a\to 0}\left[\frac{a^2}{A^2}\left(1 - (1-a)e^A\right)\right]^s = -\lim_{a\to 0}\left[\frac{a^2}{A^2}\left(1 - e^A\right)\right]^s \sim \left[\frac{\left(e^A - 1\right)}{A^2}a^2\right]^s. \tag{28}$$

Comparison between Eqs. (22) and (28) allows to make the following conclusion. If w is chosen to be A/a then the asymptotics of variances, as $a \to 0$, are the same in the densities set of $p_1(x_i), p_2(x_i)$. In the numerical simulation the relative accuracy of $\approx 8,68 \cdot 10^{-3}$ is reached as $N = 9 \cdot 10^6, \ \ s = 10, \omega = A/a, a \in [0,01; 0,001]$.

Let us turn now to the integral

$$I(f) = \int\limits_{[0,1]^{10}} f(x_1, x_2, \ldots, x_{10})\, dx_1 dx_2 \ldots dx_{10}, \tag{29}$$

where

$$f(x_1, x_2, \ldots, x_{10}) = \begin{cases} 1, & 0 \leq x_i \leq 1/4, \\ 0, & \text{otherwise}. \end{cases} \tag{30}$$

We put $p(x) \equiv 1$; then $I(f) = I(f/p) = I(f^2/p) = (1/4)^{10} \approx 9,5367 \cdot 10^{-7}$. For $N = 640000$ all realizations are turned out to be equal to zero, i.e., $q_{640000}(x) = 0$ as $0 \leq x_i \leq 1/4$. In this case, we have

$$\int_{[0,1]^{10}} p(x) \frac{f(x)}{p(x)} dx - \int_{[0,1]^{10}} q_{640000} \frac{f(x)}{p(x)} dx = (1/4)^{10} - 0 = (1/4)^{10}. \tag{31}$$

In accordance with N, the realizations numbers of η_i are turned out to be equal to 1 as $N = 810000$; equal to 3 as $N = 4000000$; equal to 15 as $N = 16000000$.

We now take $p(x_i) = (\omega + 1)(1 - x_i)^{\omega}$ in the unit cube $[0, 1]^{10}$ $(\omega > 1)$. Such choice provides the gross realizations of points in $[0, 1/4]^{10}$ and as consequence, we get benefit in quality of random values (simultaneously, we have decrease of the estimation variance, and as consequence decrease of the statistical error with the 3σ rule.) **Table 5** shows the calculations results for $N = 9000000$. Note that $\hat{\sigma}$ reaches the minimum as $\omega = 5$. In this case, we have $\hat{p}(x) = 6(1 - x)^5$, $I(f/\hat{p}) = I(f^2/\hat{p}) \approx 3,49 \cdot 10^{-11}$. Making the more detailed research for both $\omega = 5$ and the theoretical value $\sigma \approx 5,83 \cdot 10^{-6}$ we get the results represented in **Table 6**. If the function

w	$\hat{I}$	rule "3σ"
3	$9,53 \cdot 10^{-7}$	$8,61 \cdot 10^{-9}$
4	$9,58 \cdot 10^{-7}$	$6,36 \cdot 10^{-9}$
5	$9,55 \cdot 10^{-7}$	$5,82 \cdot 10^{-9}$
6	$9,55 \cdot 10^{-7}$	$6,49 \cdot 10^{-9}$
7	$9,51 \cdot 10^{-7}$	$8,06 \cdot 10^{-9}$

Table 5.
The results of numerical calculations for the integral Eq. (29) at various ω values.

N	$\hat{I}$	$\hat{\sigma}$
640,000	$9,61 \cdot 10^{-7}$	$6,15 \cdot 10^{-6}$
810,000	$9,56 \cdot 10^{-7}$	$6,00 \cdot 10^{-6}$
1,000,000	$9,51 \cdot 10^{-7}$	$5,91 \cdot 10^{-6}$
4,000,000	$9,52 \cdot 10^{-7}$	$5,74 \cdot 10^{-6}$
16,000,000	$9,55 \cdot 10^{-7}$	$5,84 \cdot 10^{-6}$

Table 6.
The results of numerical calculations for the integral Eq. (29) at $\omega = 5$.

$f(x_1, x_2, \dots, x_{10})$ close to some constant in $[0, 1/4]^{10}$ and small out of this interval then we can advise to use $\hat{p} = 6(1-x)^5$ to calculate the integral in $[0, 1]^{10}$.

5. Moments calculation

We are now concerned with the following issue: to find the kth moments of a random value τ with the distribution density $p(x)$:

$$\mathcal{E}\tau^k = \int_a^b x^k p(x)\,dx. \tag{32}$$

In fact we have realizations of the random value ξ with a distribution density $q(x)$. With $p(x)$ replaced by $q(x)$ in Eq. (32) we get an error

$$\int_a^b x^k p(x)\,dx - \int_a^b x^k q(x)\,dx = \int_a^b x^k [p(x) - q(x)]\,dx. \tag{33}$$

Suppose $b = \infty$ and ξ_{max} are the maximum value of the random variable over the all realizations for fixed N; then value of ξ_{max} gives shift $\int_{\xi_{max}}^{\infty} x^k p(x)\,dx$ that increases both monotonically and without limit. The condition $q(x) = 0$ as $x > \xi_{max}$ determines the lower limit of the last integral.

Many solutions of the boundary-value problems for the elliptic and parabolic Equations [4, 5] have a form of the expectations for the random value moments. Meaning of these expectations is the first exit time of the Wiener process trajectories to the domain boundary.

Let a domain be the three-dimensional ball with the radius $r = 1$ and the Wiener trajectories start from the ball center; then a function of distribution of the first exit time for the Wiener trajectory is, in particular, given by [5].

$$F(t) = 1 + 2\sum_{k=1}^{\infty} (-1)^k \exp\left(-k^2\pi^2 t/2\right), \quad t \in [0, +\infty). \tag{34}$$

From the above, we obtain the distribution density:

$$p(t) = 2\sum_{k=1}^{\infty} (-1)^{k+1} \mu k^2 \exp\left(-\mu k^2 t\right), \quad \mu = \pi^2/2. \tag{35}$$

Assuming τ is distributed with this density and calculating the expectation of the kth moment we get

$$\mathcal{E}\tau^k = \int_0^{\infty} t^k p(t)\,dt. \tag{36}$$

In **Table 7**, we put the calculations results for $N = 1000000$. The kth moment expectation can be represented in a form.

$$\mathcal{E}\tau^k = \int_0^\infty q(x)t^k \frac{p(t)}{q(t)}\,dt, \qquad (37)$$

where $q(t)$ is some density in $[0, \infty)$. Taking $q(t) = \lambda \exp(-\lambda t)$, for $\lambda = \pi^2/2$ we get that the number of realizations $\xi_i > 1$ will be almost twice as small as in the case of the modeling with the original $p(t)$. In this situation we should obtain degradation of the estimation for the high moments. The calculations results with $q(t)$ for $N = 1000000$ are represented in **Table 8**. However, in realizations at a computer we get the obvious

Moment	Simulation	Theory
1	$3,304 \cdot 10^{-1}$	$3,333 \cdot 10^{-1}$
2	$1,553 \cdot 10^{-1}$	$1,556 \cdot 10^{-1}$
3	$9,848 \cdot 10^{-2}$	$9,841 \cdot 10^{-2}$
4	$8,076 \cdot 10^{-2}$	$8,063 \cdot 10^{-2}$
5	$8,291 \cdot 10^{-2}$	$8,193 \cdot 10^{-2}$
6	$9,843 \cdot 10^{-2}$	$9,969 \cdot 10^{-2}$
7	$1,319 \cdot 10^{-1}$	$1,414 \cdot 10^{-1}$
8	$2,070 \cdot 10^{-1}$	$2,293 \cdot 10^{-1}$
9	$4,518 \cdot 10^{-1}$	$4,182 \cdot 10^{-1}$
10	$7,286 \cdot 10^{-1}$	$8,474 \cdot 10^{-1}$
11	$9,021 \cdot 10^{2}$	$4,251 \cdot 10^{3}$
12	$1,183 \cdot 10^{3}$	$1,637 \cdot 10^{4}$
13	$8,389 \cdot 10^{3}$	$6,634 \cdot 10^{4}$

Table 7.
The results of numerical calculations for the moments by the first way.

Moment	Simulation
5	$8,188 \cdot 10^{-2}$
6	$1,013 \cdot 10^{-1}$
7	$1,468 \cdot 10^{-1}$
8	$2,247 \cdot 10^{-1}$
9	$3,950 \cdot 10^{-1}$
10	$8,283 \cdot 10^{-1}$
18	$2,833 \cdot 10^{3}$
19	$1,056 \cdot 10^{4}$
20	$5,118 \cdot 10^{4}$

Table 8.
The results of numerical calculations for the moments by the second way.

improvement in quality of the moments estimation for all k from 5 to 20. Consider the choice of modeling strategy with regards to the variance. Suppose ξ and η be estimations of the statistical modeling for a value J, i.e., $\mathcal{E}\xi = \mathcal{E}\eta = J$ with the variances of $\sigma_1^2(\xi), \sigma_2^2(\eta)$ and the realizations of $\xi = (\xi_1 + \ldots + \xi_N)/N, \eta = (\eta_1 + \ldots + \eta_N)/N$. It would seem that for $\sigma_1(\xi) < \sigma_2(\eta)$ the real estimation of ξ will be occurred close to the origin value of J. But this statement does not need to be always true. Without loss of generality it can believed that $J = 0$. Additionally, if N is large enough then ξ, η are chosen be normal random variables with $N(0, \sigma_1)$ and $N(0, \sigma_2)$, respectively. The following theorem holds.

Proposition 6. Let ξ, η be normal random variables, and $\xi \sim N(0, \sigma_1), \eta \sim N(0, \sigma_2)$ then $P(\xi| > |\eta|) = \frac{2}{\pi} \arctan \frac{\sigma_1}{\sigma_2}$.

Proof:

$$P|\xi > |\eta| = \frac{1}{\sigma_1\sqrt{2\pi}} \int_{-\infty}^{0} e^{-\frac{y^2}{2\sigma_1^2}} dy \left\{ \frac{1}{\sigma_2\sqrt{2\pi}} \int_{0}^{y} e^{-\frac{x^2}{2\sigma_2^2}} dx + \frac{1}{\sigma_2\sqrt{2\pi}} \int_{0}^{|y|} e^{-\frac{x^2}{2\sigma_2^2}} dx \right\} +$$

$$+ \frac{1}{\sigma_1\sqrt{2\pi}} \int_{0}^{\infty} e^{-\frac{y^2}{2\sigma_1^2}} dy \left\{ \frac{1}{\sigma_2\sqrt{2\pi}} \int_{0}^{y} e^{-\frac{x^2}{2\sigma_2^2}} dx + \frac{1}{\sigma_2\sqrt{2\pi}} \int_{0}^{|y|} e^{-\frac{x^2}{2\sigma_2^2}} dx \right\} =$$

$$= \frac{2}{\sigma_1\sqrt{2\pi}} \int_{0}^{\infty} e^{-\frac{y^2}{2\sigma_1^2}} dy \left\{ \frac{2}{\sigma_2\sqrt{2\pi}} \int_{0}^{y} e^{-\frac{x^2}{2\sigma_2^2}} dx \right\} = \frac{4}{2\pi\sigma_1\sigma_2} \int_{0}^{\infty} e^{-\frac{y^2}{2\sigma_1^2}} dy \cdot \int_{0}^{y} e^{-\frac{x^2}{2\sigma_2^2}} dx.$$

Using Taylor expansion

$$e^{-\frac{y^2}{2\sigma_2^2}} = \sum_{n=0}^{\infty} (-1)^n \frac{y^{2n}}{2^n \sigma_2^{2n} n!},$$

we get

$$\int_{0}^{y} \sum_{n=0}^{\infty} (-1)^n \frac{x^{2n}}{2^n \sigma_2^{2n} n!} dx = \sum_{n=0}^{\infty} (-1)^n \frac{y^{2n+1}}{(2n+1) 2^n \sigma_2^{2n} n!},$$

$$\int_{0}^{\infty} \sum_{n=0}^{\infty} (-1)^n \frac{y^{2n+1}}{(2n+1) 2^n \sigma_2^{2n} n!} e^{-\frac{y^2}{2\sigma_1^2}} dy = \sum_{n=0}^{\infty} (-1)^n \frac{1}{(2n+1) 2^n \sigma_2^{2n} n!} \int_{0}^{\infty} y^{2n+1} e^{-\frac{y^2}{2\sigma_1^2}} dy =$$

$$= \sum_{n=0}^{\infty} (-1)^n \frac{1}{(2n+1) 2^n \sigma_2^{2n} n!} \cdot \frac{2^{n+1} \sigma_1^{2n+2} n!}{2}. \quad (*)$$

Note that the last equality is obtained with the help of the formula:

$$\int_{0}^{\infty} x^{2n+1} e^{-px^2} dx = \frac{n!}{2p^{n+1}}, \quad p > 0.$$

In our case, p is $\frac{1}{2\sigma_1^2}$. We continue the equalities chain which is broken at (*):

Moment	Simulation
1	$3,297 \cdot 10^{-1}$
2	$1,556 \cdot 10^{-1}$
3	$9,842 \cdot 10^{-2}$
4	$8,066 \cdot 10^{-2}$
5	$8,194 \cdot 10^{-2}$
6	$9,968 \cdot 10^{-2}$
7	$1,414 \cdot 10^{-1}$
8	$2,293 \cdot 10^{-1}$
9	$4,181 \cdot 10^{-1}$
10	$8,474 \cdot 10^{-1}$
18	$4,251 \cdot 10^{3}$
19	$1,637 \cdot 10^{4}$
20	$6,634 \cdot 10^{4}$

Table 9.
The results of numerical calculations for the moments by the third way.

$$(*) = \sum_{n=0}^{\infty} (-1)^n \frac{1}{(2n+1)} \frac{\sigma_1^{2n+2}}{\sigma_2^{2n}} = \sum_{n=0}^{\infty} (-1)^n \frac{\sigma_1^2}{(2n+1)} \left(\frac{\sigma_1}{\sigma_2}\right)^{2n} =$$

$$= \frac{2}{\pi} \sum_{n=0}^{\infty} (-1)^n \frac{(\sigma_1/\sigma_2)^{2n+1}}{2n+1} = \frac{2}{\pi} \arctan \frac{\sigma_1}{\sigma_2}.$$

For the k-moment calculation we take

$$q_k(t) = \frac{\lambda^{k+1} t^k e^{-\lambda t}}{k!}, \lambda = \frac{\pi^2}{2}$$

and get the results shown in **Table 9**.

6. Integral from the regression function

Now, we consider the issue of calculation of an integral

$$\int_D f(x) dx, \tag{38}$$

where the function $f(x)$ has no an analytical expression. Suppose there exists a random variable $\xi(x, w)$ such that its expectation is equals to $\mathcal{E}\xi(x, w) = f(x)$ for some fixed x. The random variable $\xi(x, w)$ may be realized neither as result of the physical measurements or some calculations (e.g., using the modeling statistical method). In this case the optimal density is given by [6].

$$p(x) = \frac{f(x)}{\sqrt{d(x) + \lambda}}, \tag{39}$$

where $d(x)$ is the variance of the random variable $\xi(x, w)$. Note that one should use the optimal density from [1] if complexity in calculations (experimental measurements) is much different from each other for any x. We determine the parameter λ from the condition $\int\limits_D f(x)/\sqrt{d(x) + \lambda} dx = 1$. Really in practice, we find a priori or a posteriori approaches to both $f(x)$ and $d(x)$. By $\bar{f}(x)$ and $\bar{d}(x)$ denote these approaches. Then the approach to the optimal $p(x)$ will look like

$$\bar{p}(x) = \frac{\bar{f}(x)}{\sqrt{\bar{d}(x) + \bar{\lambda}}} \text{ and } \int\limits_D \frac{\bar{f}(x)}{\sqrt{\bar{d}(x) + \bar{\lambda}}} dx = 1. \tag{40}$$

The parameter λ is often turned out to be find enough complicity [6]. If the domain D is the interval $[0, H]$ for small H then it is suppose to use the quasioptimal density $\bar{p}(x)$.

Example 3. We now consider the following issue: Suppose $f(x) = x$, $d(x) = 1/x$, $D = [0, H]$. The optimal density is given by

$$p(x) = \frac{x}{\sqrt{1/x + \lambda}} = \frac{x\sqrt{x}}{\sqrt{\lambda\sqrt{x} + 1}} \sim c \cdot x^{3/2}. \tag{41}$$

We take the quasioptimal density in the form $\bar{p}(x) = 5H^{5/2}x^{3/2}/2$. In this case, for $p(x) \equiv 1$ the estimation variance of the random value $\eta = f(x, w)/p(x)$:

$$\mathbf{var}\eta = \int\limits_0^H d(x)p(x)dx + \int\limits_0^1 \frac{f^2}{p(x)} dx - I^2 \tag{42}$$

is equals to ∞. Taking $p(x) = 2x/H^2$ we get $\mathbf{var}\eta = 2/H$. But if the function $f(x)$ was precisely known for the same density $p(x)$ then $\mathbf{var}\eta = 0$. If we choose the quasioptimal density $\bar{p}(x) = 5H^{5/2}x^{3/2}/2$ then the estimation variance of η is equals to $17H^4/12 + 4/(15H)$. For $H \to 0$ the variance behaves approximately as $4/(15H)$. It is much the better than $2/H$. For $H = 1$ the estimation variance with the density $p(x) = 2x/H^2$ is equals to 2, and the estimation variance with the quasioptimal density $\bar{p}(x)$ is equals to 101/60.

Suppose we practically realize calculation of the integral Eq. (38) with $d(x) = 1/x$; then one should discard the interval $[0, \delta]$ and to calculate $\int_\delta^H f(x)dx$ because of the values $|\xi(x, w)|$ can be the intolerably large. Also one should replace $f(x)$ by $\hat{f}(x)$:

$$\hat{f}(x) = \begin{cases} 0, & 0 \le x \le \delta, \\ x, & \delta < x \le H. \end{cases} \tag{43}$$

The shift is $\int\limits_0^\delta x dx = \delta^2/2$ and choosing $\delta \sim 1/\sqrt[4]{N}$ we get the total error $\delta^2/2 + 3\sigma/\sqrt{N}$ of oder $O(1/\sqrt{N})$.

In applications the estimation variance for the integral functionals (e.g., field flow calculation neither across the arc or the surface) from the solutions of the boundary-value problems for both the linear [1] or nonlinear [2] elliptic equations is of interest. For the above variance is $d(x) \sim B/x^2,\ f(x) \approx a_0 + a_1x + a_2x^2 + \ldots$, where x is the distance to the domain boundary. Suppose $f(x) \approx a_1x + a_2x^2 + \ldots$; then the optimal density is given by

$$p(x) = \frac{a_1x + a_2x^2 + \ldots}{\sqrt{B/x^2 + \lambda}}. \tag{44}$$

The quasioptimal density has the form $\overline{p}(x) = 3x^2/H^3$ for small H in $[0, H]$. In applications, this case is of our main interest. Taking $\delta = 1/\sqrt[4]{N}$ like in the Example 3 we get the asymptotics of decrease for the total error as $O(1/\sqrt{N})$.

Suppose $d(x) \sim B/x^2,\ f(x) \approx a_0 + a_1x + a_2x^2 + \ldots$, and $a_0 \neq 0$ then there is no density kind of $\overline{p}(x) = (w+1)x^w, x \in [0, H]$ with the finite variance. The density $p(x) = |f(x)|/\sqrt{d(x) + \lambda}$ will be give the estimation with the infinity variance. Instead of calculation of the integral $\int_0^H f(x)dx$ we will be calculate the integral $\int_\delta^H f(x)dx$. For this integral we already can choice the quasioptimal density with the finite variance of the estimation: $\overline{p}(x) = 2x/(H^2 - \delta^2)$. For $\delta \sim O(\ln N/\sqrt{N})$ the total error will have the asymptotics $O(\ln N/\sqrt{N})$.

Example 4. Suppose $d(x) = 1/x^2,\ f(x) = 1,\ H = 1$; then the asymptotics of the variance with the quasioptimal density has kind of $(-2,5 \cdot \ln \delta)$.

In conditions of Example 4, choice of the optimal density in the form

$$p(x) = \frac{x}{\sqrt{1-\delta^2}\sqrt{1-x^2}}, \quad x \in [\delta, 1] \tag{45}$$

yields the following result: the estimation variance will have asymptotics $(-2 \cdot \ln \delta)$ for $\delta \to 0$.

Remark. If we know that a value of $f(x)$ in the interval $[0, \delta]$ close to the number f_0, then in Eq. (43) to use

$$\hat{f}(x) = \begin{cases} f_0, & 0 \le x \le \delta, \\ x, & \delta < x \le H, \end{cases} \tag{46}$$

more efficiently and also to take $\int_0^H f(x)dx \approx f_0\delta + \int_\delta^H f(x)dx$.

7. Conclusion

In the paper we describe the sufficient conditions of the stable calculations for the multidimensional integrals by the Monte Carlo method. We get the results of numerous numerical computations using the mt19937 pseudorandom number generator. The article results can be also useful in the practical solution of the boundary value problem, for both the elliptic and parabolic equations. The earlier suggested approach to the optimal choice of the density [1, 6] often needs to solve a complicated secondary task. In the paper we suggest the approach to choice of the quasioptimal densities that is of considerable interest in applied problems solution.

Author details

Alexander A. Kronberg*† and Tatiana K. Kronberg†
Independent Scientists, Dubna, Russia

*Address all correspondence to: kronberg.alexander@gmail.com

† These authors contributed equally.

References

[1] Kronberg AA. On the numerical finding of some functionals on solving of Laplace's equation. Izv. Vuz. Matematika. 1987;**31**:30-36

[2] Kronberg AA. Algorithms for solving elliptic boundary value problems. UUSR Computational Mathematics and Mathematical Physics. 1982;**22**:122-131. DOI: 10.1016/0041-5553(82)90103-3

[3] Matsumoto M, Nishimura T. Mersenne twister: A 623-dimensionally equidistributed uniform preudorandom number. ACM Transactions on Modeling and Computer Simulation (TOMACS). 1998;**8**:G1-G25. DOI: 10.1145/272991.272995

[4] Dynkin EB. Markov Processes. Vol. 2. Berlin: Springer-Verlag; 1965. p. 274

[5] Haji-Sheiki A, Sparrow EM. The solution of heat conduction problems by probability methods. Transmission ASME Series C: Journal of Heat Transfer. 1967;**89**:121-131. DOI: 10.1115/1.3614330

[6] Kronberg AA. On the numerical solution of Laplace's equation in unbounded domain. Izv. Vuz. Matematika. 1984;**28**:45-48

Chapter 4

Some Results on the Non-Homogeneous Hofmann Process

Gerson Yahir Palomino Velandia and José Alfredo Jiménez Moscoso

Abstract

The classical counting processes (Poisson and negative binomial) are the most traditional discrete counting processes (*DCPs*); however, these are based on a set of rigid assumptions. We consider a non-homogeneous counting process (which we name non-homogeneous Hofmann process – *NHP*) that can generate the classical counting processes (*CCPs*) as special cases, and also allows modeling counting processes for event history data, which usually exhibit under- or over-dispersion. We present some results of this process that will allow us to use it in other areas and establish both the probability mass function (*pmf*) and the cumulative distribution function (*cdf*) using transition intensities. This counting process (*CP*) will allow other researchers to work on modelling the *CP*, where data dispersion exists in an efficient and more flexible way.

Keywords: mixed Poisson Process, Hofmann process, variance-to-mean ratio, transition intensity

1. Introduction

In ref. [1], Hofmann introduced a new class of infinitely divisible mixed Poisson process (*MPP*), this broader class of *CP* allows obtaining other *CCP* by simply modifying or choosing its parameters, as well as Poisson, negative binomial, Poisson-Pascal among other distributions (see [2]). The family of distributions defined by Hofmann has been used in many types of applications of modelling and simulation studies that include topics such as accident models [3].

In this chapter, we analysed the event of number process $\{N(t), t \geq 0\}$ and used a broader *CP,* which is based on the Hofmann process. The appeal of this *CP* is that, analogous to the family of frequency distributions, it allows to generate several known *CP*. Through an *NHP,* we can generate the following as special cases: the Poisson counting process (*PCP*), the negative binomial counting process (*NBCP*) and the Poisson-Pascal process among other *CCPs*, and this allows us to obtain models for *CP* with under- or over-dispersion. The *NHP* was introduced by Hofmann [1] and has been used by other researchers [3–5]. Some properties of the *NHP* found by Walhin

[2] are presented in this chapter, and we used the transition intensities to describe additional properties of the *NHP*.

The objective of this chapter is to present a unified view of related results on the *NHP*. The chapter is organised as follows: in Section 2, we present the *NHP*; in Section 3, we present some statistical properties, such as *pmf* and probability generating function (pgf), and formulas for the mean and variance are derived; in Section 4, we present various approaches for the *NHP* using *CCP*; in Section 5, we present other properties for *NHP*; finally, conclusions are presented.

2. Basic concepts of the *NHP*

Let us take $N(t)$ as the number of events that occurs in the time interval $(0, t]$ with $t > 0$ and $N(0) = 0$. The probability of n events occurring in this time interval is denoted by

$$P_n(t) = P[N(t) = n], \qquad n = 0, 1, 2, \dots \tag{1}$$

According to Dubourdieu [6], an *MPP* $\{N(t) : t \geq 0\}$ is a *PCP* with rate Λ, where the non-negative random variable Λ is called a structure variable. The *MPP* has been studied by several authors [7–9].

When Λ is a continuous random variable with probability density function (*pdf*), $f(\lambda)$, we can find probability by

$$\begin{aligned} \underbrace{\mathbb{E}[P[N(t) = n|\Lambda]]}_{P[N(t)=n]} &= \int_0^\infty P[N(t) = n|\Lambda = \lambda] f(\lambda) d\lambda \\ &= \int_0^\infty e^{-\lambda t} \frac{(\lambda t)^n}{n!} f(\lambda) d\lambda. \end{aligned} \tag{2}$$

For $n = 0$ and $t > 0$ we have

$$P_0(t) = \int_0^\infty e^{-\lambda t} f(\lambda) d\lambda, \tag{3}$$

The higher order derivatives of the last expression with respect to t are

$$P_0^{(n)}(t) = \frac{d^n}{dt^n} P_0(t) = (-1)^n \int_0^\infty \lambda^n e^{-\lambda t} f(\lambda) d\lambda. \tag{4}$$

By substituting (4) into (2) we get

$$P_n(t) = \frac{t^n}{n!}\left[(-1)^n P_0^{(n)}(t)\right], \qquad n \geq 1 \tag{5}$$

The expressions (3) and (5) characterize an *MPP* with a continuous structure variable Λ. According to Hofmann [1], for the construction of examples, a special structure function is generally assumed, and from this the *pmf* is calculated by (3), (5). In most cases, this leads to formally complicated expressions. In ref. [1], Hofmann presents a *CP* called Hofmann process as an option to model the event number process given by (2) and whose general expression for (3) is as follows:

$$P_0(t) = \exp\{-\theta(t)\} \qquad \theta(t) = \int_0^t \lambda(\tau; a)d\tau \tag{6}$$

where $P_0(t)$ is a completely monotonic function[1]. And $\lambda(\tau; a)$ is a function of three parameters: $a \geq 0$, $q > 0$ and $\kappa \geq 0$, which is a function infinitely divisible and given by

$$\lambda(\tau; a) = \frac{q}{(1+\kappa\tau)^a} \qquad \forall \tau > 0. \tag{7}$$

Although $\lambda(\tau; a)$ depends on three parameters, we use this notation given that the parameter a provides various *CCPs*. We denote the *NHP* by $\mathcal{H}(a, q, \kappa)$, if the *pmf* of $N(t)$ satisfies the expressions (5) and (6).

Using the expression (7), we get by integrating that

$$\theta(t) = \begin{cases} \ln\left[(1+\kappa t)^{q/\kappa}\right] & \text{if} \quad a = 1 \\ \dfrac{q}{\kappa(1-a)}\left[(1+\kappa t)^{1-a} - 1\right] & \text{if} \quad a \neq 1 \end{cases} \tag{8}$$

By substituting (8) into (6)

$$P_0(t) = \begin{cases} (1+\kappa t)^{-\frac{q}{\kappa}} & \text{if} \quad a = 1 \\ \exp\left\{-\dfrac{q}{\kappa \cdot (1-a)}\left[(1+\kappa t)^{1-a} - 1\right]\right\} & \text{if} \quad a \neq 1 \end{cases} \tag{9}$$

Remark 1.1: If in the expression (9) for $a = 1$ we take the limit as $\kappa \to 0$, we have:

$$\lim_{\kappa \to 0} (1+\kappa t)^{-\frac{q}{\kappa}} = e^{-qt}, \tag{10}$$

and the last expression agrees with the adequate $P_0(t)$ of a *PCP* with rate qt.

3. Basic properties of the *NHP*

Theorem 1.2: Let $N(t)$ be an NHP then

i. The pgf of the process is given by

$$G_N(z; t) = \begin{cases} (1+\kappa(1-z)t)^{-q/\kappa} & \text{if} \quad a = 1 \\ \exp\left\{-\dfrac{q}{\kappa(1-a)}\left[(1+\kappa(1-z)t)^{1-a} - 1\right]\right\} & \text{if} \quad a \neq 1 \end{cases} \tag{11}$$

---

[1] We say that a function $g(t)$ with $t \in \mathbb{R}^+$ is completely monotonic if it has derivatives $g^{(n)}(t)$ for all $n \in \mathbb{N}$ and its derivatives have alternating signs, i.e., if $(-1)^n g^{(n)}(t) \geq 0$, $t > 0$.

Note that $G_N(z;t) = P_0((1-z)t)$ with $0 \leq z < 1$.

ii. The *pmf* of $N(t)$, for t fixed, satisfies the following recursive formula:

$$P_{n+1}(t) = \frac{t\lambda(t;a)}{n+1}\sum_{i=0}^{n}\binom{a+i-1}{i}\left(\frac{\kappa t}{1+\kappa t}\right)^{i} P_{n-i}(t) \tag{12}$$

where $P_0(t) = G_N(0;t)$ is given by (9) and

$$P_0^{(n+1)}(t) = \lambda(t;a)\sum_{j=0}^{n}\binom{n}{j}(-1)^{j+1}\frac{\Gamma(a+j)}{\Gamma(a)}\left(\frac{\kappa}{1+\kappa t}\right)^{j} P_0^{(n-j)}(t)$$

iii. If $a = 1$ the $P_n(t)$ satisfies the recurrence relation

$$\frac{P_{n+1}(t)}{P_n(t)} = \frac{-t}{n+1}\frac{P_0^{(n+1)}(t)}{P_0^{(n)}(t)} = \frac{q+\kappa n}{1+\kappa t}\frac{t}{n+1}. \tag{13}$$

iv. The process $N(t)$ has a mean and variance given by

$$\mathbb{E}[N(t)] = qt \qquad \text{and} \qquad Var[N(t)] = (1+a\kappa t)\mathbb{E}[N(t)] \tag{14}$$

Proof:
See details in [2] or [10].
Note that from (14) we have that if $q \neq 0$ then:

$$\lim_{t\to\infty}\frac{\mathbb{E}[N(t)]}{t} = q. \tag{15}$$

It is possible from (14) to calculate the measure based on the variance-to-mean ratio (VMR) introduced by [11]:

$$ID(t) = \frac{Var[N(t)]}{\mathbb{E}[N(t)]} = 1 + a\kappa t. \tag{16}$$

As $ID(t) > 1$, then using the criterion of the VMR, we have that the *NHP* is an over-dispersed *CP* and hence is an option for modelling over-dispersion in count data.
Using the expression (11), in **Table 1**, we present the functions for qt and κt that allow to obtain some *CP*. We consider the *CCPs* studied in [10], which are special cases of *NHP* when $a = 1$ since this reduces to the Panjer counting process (see [12]). In addition, we consider other processes, such as the Neyman Type A process introduced by [13], the Poisson Pascal process introduced by [14] and the Pólya-Aeppli process introduced by [15].

3.1 *NHP* is infinitely divisible

The following relationships are identical to those of [16] which characterize infinitely divisible *pmf*:

	Counting process	$P_0[(1-z)t]$	Functions	
			qt	κt
Classical ($a = 1$)	Poisson	$\exp\{-(1-z)\gamma t\}, \quad \kappa \to 0$	γt	0
	Negative binomial (or Pólya)	$\left[\frac{\delta}{\delta+(1-z)t}\right]^{\gamma}, \quad \delta > 0$	$\frac{\gamma}{\delta}t$	$\frac{t}{\delta}$
	Geometric	$\frac{\delta}{\delta+(1-z)t}$	$\frac{t}{\delta}$	$\frac{t}{\delta}$
Other ($a > 1$)	Neyman Type A	$\exp\{\gamma[\exp\{(z-1)\delta t\}-1]\}, \quad a \to \infty$	$\gamma\delta t$	$\frac{\delta t}{a-1}$
	Poisson-Pascal	$\exp\left\{\gamma\left[(1+(1-z)\delta t)^{-(a-1)}-1\right]\right\}$	$(a-1)\gamma\delta t$	δt
	Pólya-Aeppli	$\exp\left\{\frac{-(1-z)\gamma t}{1-\left[1-(1+\delta t)^{-1}\right]z}\right\}, \quad a = 2$	$(1+\delta t)\gamma t$	δt
Source: own elaboration				

Table 1.
Functions qt and κt for some CCPs.

Theorem 1.3: The *pmf* $\{P_n(t)\}$ with $P_0(t) > 0$ is infinitely divisible if and only if satisfies that

$$(n+1)P_{n+1}(t) = \sum_{i=0}^{n} r_i(t)P_{n-i}(t) \qquad \text{for } t \text{ fixed.}$$

where the quantities $r_n(t)$ with $n \in \mathbb{Z}^+$ are nonnegative.
Proof: See details in [16].
Corollary 1.3.1: The *pmf* $\{P_n(t)\}$ of the *NHP* is infinitely divisible.
Proof:
By multiplying (12) by $(n+1)$ we get

$$(n+1)P_{n+1}(t) = \sum_{i=0}^{n} t\lambda(t;a)\binom{a+i-1}{i}\left(\frac{\kappa t}{1+\kappa t}\right)^{i} P_{n-i}(t).$$

We denote

$$r_i(t;a) = qt\binom{a+i-1}{i}\frac{(\kappa t)^i}{(1+\kappa t)^{a+i}} \qquad i = 0, 1, \dots, n. \tag{17}$$

Note that $r_i(t;a) \geq 0$, which allows to conclude that $P_n(t)$ is infinitely divisible.
The following relationship is given by [17]: all log-convex distributions are infinitely divisible but not all log-concave distributions are infinitely divisible.
Theorem 1.4: Let $N(t)$ be an infinitely divisible $\mathbb{Z}^+$-valued random variable with pmf $P_n(t)$. Then

$$\mathbb{E}[N(t)] = \sum_{i=0}^{\infty} r_i(t;a) \tag{18}$$

Proof:
We know that the expectation of $N(t)$ it is given by

$$\mathbb{E}[N(t)] = \sum_{n=1}^{\infty} nP_n(t) = \sum_{m=0}^{\infty}(m+1)P_{m+1}(t)$$
$$= \sum_{m=0}^{\infty}\sum_{i=0}^{m} r_i(t;a)P_{m-i}(t)$$

Now, by interchanging the order of summation, we get

$$\mathbb{E}[N(t)] = \sum_{i=0}^{\infty}\sum_{m=i}^{\infty} r_i(t;a)P_{m-i}(t) = \sum_{i=0}^{\infty} r_i(t;a)\sum_{m=i}^{\infty} P_{m-i}(t)$$
$$\underset{j=m-i}{=} \sum_{i=0}^{\infty} r_i(t;a)\sum_{j=0}^{\infty} P_j(t) = \sum_{i=0}^{\infty} r_i(t;a).$$

which completes the proof.

4. *NHP* in terms of *CCPs*

In this section, we present various approaches for the *NHP* using *CCP*.

4.1 *NHP* as a non-homogeneous pure birth process

We use logarithmic differentiation to find the derivative of (5) and we get

$$\frac{P_{n'}(t)}{P_n(t)} = \frac{n}{t} + \frac{P_0^{(n+1)}(t)}{P_0^{(n)}(t)}$$

Then

$$P_{n'}(t) = \frac{n}{t}P_n(t) + \frac{P_0^{(n+1)}(t)}{P_0^{(n)}(t)}P_n(t) \tag{19}$$

From (5), we obtain

$$\frac{n}{t}P_n(t) = -\frac{(-1)^{n-1}}{(n-1)!}t^{n-1}P_0^{(n)}(t) = \frac{(-1)^{n-1}}{(n-1)!}t^{n-1}P_0^{(n)}(t)\left(-\frac{P_0^{(n-1)}(t)}{P_0^{(n-1)}(t)}\right)$$
$$= -\frac{P_0^{(n)}(t)}{P_0^{(n-1)}(t)}P_{n-1}(t)$$

By substituting in (19), we have

$$P_{n'}(t) = \left(-\frac{P_0^{(n)}(t)}{P_0^{(n-1)}(t)}\right)P_{n-1}(t) - \left(-\frac{P_0^{(n+1)}(t)}{P_0^{(n)}(t)}\right)P_n(t). \tag{20}$$

We denote

$$\lambda_n(t;a) = -\frac{P_0^{(n+1)}(t)}{P_0^{(n)}(t)} = -\frac{d}{dt}\ln\left[(-1)^n P_0^{(n)}(t)\right]. \tag{21}$$

In ref. [18], Lundberg shows that this corresponds to the transition intensities. Then from (20) and (21), we can derive the following system of Kolmogorov differential equations that must be satisfied by the *NHP*:

$$\begin{aligned} P_0'(t) &= -\lambda_0(t;a)P_0(t) \\ P_n'(t) &= \lambda_{n-1}(t;a)P_{n-1}(t) - \lambda_n(t;a)P_n(t) \qquad \text{for} \qquad n \geq 1. \end{aligned} \tag{22}$$

By notation, we denote $\lambda_0(t;a) = \theta'(t) = \frac{q}{(1+\kappa t)^a}$. With initial conditions

$$P_0(0) = 1 \quad \text{and} \quad P_n(0) = 0 \quad \forall n \geq 1 \tag{23}$$

Using the method given in ref. [18], we find that the solution of (22) is given by

$$P_n(t) = \int_0^t \lambda_{n-1}(\tau;a)P_{n-1}(\tau)\exp\left\{-\int_\tau^t \lambda_{n-1}(\nu;a)d\nu\right\}d\tau \qquad \text{for} \qquad n \geq 1.$$

From the system of equations given in (22), we have that the *NHP* is a non-homogeneous pure birth process (*NHPBP*), which agrees with the definition given by Seal in ref. [19]. So, if $N(t)$ satisfies (6), then $N(t)$ is an *NHPBP* with transition intensities given by (21).

4.2 *NHP as MPP*

The list of equivalences provided by Lundberg in ref. [18] is satisfied by the *NHP* defined in (6), which is presented in the following theorem:

Theorem 1.5: Let $N(t)$ be an *NHP* with marginal *pmf*, given by (5) and transition intensities, given by (21). Then:

i. $\lambda_n(t;a)$ satisfy $\lambda_{n+1}(t;a) = \lambda_n(t;a) - \frac{\lambda_n'(t;a)}{\lambda_n(t;a)}$ for $n = 0, 1, \ldots$

ii. $P_n(t)$ and $\lambda_n(t;a)$ satisfy the relation

$$\frac{P_n(t)}{P_{n-1}(t)} = \frac{t}{n}\lambda_{n-1}(t;a) \quad \text{for} \quad n = 1, 2, \ldots \tag{24}$$

Proof:

i. By finding the derivative of function (21) with respect to t, we obtain

$$\begin{aligned} \lambda_n'(t;a) &= -\left[\frac{P_0^{(n+2)}(t)P_0^{(n)}(t) - P_0^{(n+1)}(t)P_0^{(n+1)}(t)}{\left(P_0^{(n)}(t)\right)^2}\right] \\ &= -\frac{P_0^{(n+2)}(t)}{P_0^{(n+1)}(t)}\frac{P_0^{(n+1)}(t)}{P_0^{(n)}(t)} + \left(-\frac{P_0^{(n+1)}(t)}{P_0^{(n)}(t)}\right)^2 \\ &= -\lambda_{n+1}(t;a)\lambda_n(t;a) + [\lambda_n(t;a)]^2 \end{aligned}$$

By dividing by $\lambda_n(t;a)$, we have

$$\frac{\lambda_n'(t;a)}{\lambda_n(t;a)} = \lambda_n(t;a) - \lambda_{n+1}(t;a) \tag{25}$$

ii. By substituting (21) into (13), we get:

$$\frac{P_n(t)}{P_{n-1}(t)} = \frac{\frac{(-1)^n}{n!}t^n P_0^{(n)}(t)}{\frac{(-1)^{n-1}}{(n-1)!}t^{n-1}P_0^{(n-1)}(t)} = -\frac{t}{n}\frac{P_0^{(n)}(t)}{P_0^{(n-1)}(t)} = \frac{t}{n}\lambda_{n-1}(t;a),$$

which completes the proof. □

In ref. [7], it is proved that the above three statements are equivalent.

Corollary 1.5.1: Let $N(t)$ be an *NHP* with transition intensities given by (21), then

$$\frac{P_n(t)}{P_0(t)} = \prod_{j=1}^{n} \frac{t\lambda_{j-1}(t;a)}{j} \tag{26}$$

Proof:

Note that

$$\frac{P_n(t)}{P_0(t)} = \prod_{j=1}^{n} \frac{P_j(t)}{P_{j-1}(t)}.$$

Substituting (24) in the above expression completes the proof.

Corollary 1.5.2: Let $N(t)$ be an *NHP* with transition intensities given by (21), then

$$\prod_{j=0}^{n-1} \lambda_j(t;a) = (-1)^n \frac{P_0^{(n)}(t)}{P_0(t)} \qquad n \geq 1. \tag{27}$$

Proof:

From (21), we get

$$\prod_{j=0}^{n-1} \lambda_j(t;a) = \prod_{j=0}^{n-1}\left(-\frac{P_0^{(j+1)}(t)}{P_0^{(j)}(t)}\right) = (-1)^n \frac{P_0^{(n)}(t)}{P_0(t)}.$$

This finishes the proof of Corollary.

The following additional properties set in ref. [9] are also satisfied by *NHP*:

Proposition 1.6: Let $\{N(t); t \geq 0\}$ be an *NHP* and Λ the continuous structure variable of the MPP. Then:

1. The transition intensities are such that

$$\mathbb{E}[\Lambda | N(t) = n] = \lambda_n(t;a). \tag{28}$$

and

$$Var[\Lambda | N(t) = n] = -\lambda_n'(t;a). \tag{29}$$

2. The mean of $N(t)$ is given by

$$\mathbb{E}[N(t)] = t\mathbb{E}[\Lambda]. \tag{30}$$

3. The mean of Λ is given by

$$\mathbb{E}[\Lambda] = -P_0'(0). \tag{31}$$

Proof:
1. From (2), taking the expected value of Λ, conditioning on $N(t)$, we get

$$\mathbb{E}[\Lambda|N(t) = n] = \int_0^\infty \frac{\lambda e^{-\lambda t}(\lambda t)^n f(\lambda)}{n!P[N(t) = n]} d\lambda = \frac{n+1}{t}\frac{P_{n+1}(t)}{P_n(t)}. \tag{32}$$

By substituting (24) into (32), we have

$$\mathbb{E}[\Lambda|N(t) = n] = \lambda_n(t; a).$$

Analogously, we can show that

$$\mathbb{E}\left[\Lambda^2|N(t) = n\right] = \int_0^\infty \frac{\lambda^2 e^{-\lambda t}(\lambda t)^n f(\lambda)}{n!P[N(t) = n]} d\lambda = \frac{(n+2)(n+1)}{t^2}\frac{P_{n+2}(t)}{P_n(t)}. \tag{33}$$

By substituting (24) into (33), we have

$$\mathbb{E}\left[\Lambda^2|N(t) = n\right] = \lambda_{n+1}(t; a)\lambda_n(t; a).$$

Then the conditional variance of Λ, given that $N(t) = n$, is

$$Var[\Lambda|N(t) = n] = \lambda_{n+1}(t; a)\lambda_n(t; a) - \lambda_n^2(t; a),$$

and substituting Eq. (25) into the above yields the result.

2. We use the law of total expectation to find the expected value

$$\mathbb{E}[\Lambda] = \mathbb{E}[|\mathbb{E}(\Lambda|N(t) = n)] = \sum_{n=0}^\infty \mathbb{E}(|\Lambda|N(t) = n)P[N(t) = n]$$

$$= \sum_{n=0}^\infty \lambda_n(t; a)P_n(t)$$

By substituting (24) into the above expression, we get

$$\mathbb{E}[\Lambda] = \sum_{n=0}^\infty \frac{n+1}{t} P_{n+1}(t) = \sum_{j=0}^\infty \frac{r_j(t; a)}{t} = \frac{1}{t}\mathbb{E}[N(t)].$$

And the proof is completed.

3. The pgf of $N(t)$ is defined as

$$\begin{aligned} \underbrace{G_N(z; t)}_{} &= \sum_{n=0}^\infty z^n P_n(t) = \sum_{n=0}^\infty z^n \int_0^\infty \frac{(\lambda t)^n}{n!} e^{-\lambda t} f(\lambda) d\lambda \\ P_0[(1-z)t] &= \int_0^\infty \left[\sum_{n=0}^\infty \frac{(z\lambda t)^n}{n!}\right] e^{-\lambda t} f(\lambda) d\lambda = \int_0^\infty e^{\lambda(z-1)t} f(\lambda) d\lambda \\ &= M_\Lambda[(z-1)t]. \end{aligned} \tag{34}$$

We make $z = 0$ in the above expression and we have

$$P_0(t) = M_\Lambda(-t)$$

Now, if we differentiate both sides with respect to t, we obtain

$$P_0'(t) = -M_\Lambda'(-t)$$

We complete the proof by substituting $t = 0$ in the above expression. □

According to Walhin and Paris in ref. [20], the intensity of the stochastic process $N(t)$ in the period $[t, t+1]$ is

$$\mathbb{E}[N(t+1) - N(t)|N(t) = n] = \mathbb{E}[\Lambda|N(t) = n].$$

The moment generating function of the process will uniquely determine the distribution of the process, on comparing expression (34) with $P_0[(1-z)t]$ given for $a = 1$ and as shown in **Table 1**, we find the particular cases: the *PCP* if $\Lambda \sim \delta_\gamma(\lambda)$ (i.e. has a degenerate *cdf* at $\lambda = \gamma$), the *NBCP* if $\Lambda \sim \Gamma(\gamma, \delta)$ and the Geometric Counting Process if $\Lambda \sim \exp(\delta)$.

5. Additional properties

In this Section, we will introduce several other properties of the *NHP*.

5.1 Other expressions for $\mathbf{P_n(t)}$ in terms of $\boldsymbol{\lambda_n(t;a)}$

Theorem 1.7: Let $N(t)$ be an *NHP* with transition intensities given by (21), then

$$P_n(t) = Q_n(t) - Q_{n+1}(t) \quad \text{for} \quad n \geq 1,$$

where $Q_0(t)$ is Heaviside's step function and

$$Q_{n+1}(t) = \int_0^t \lambda_n(v;a)P_n(v)dv. \tag{35}$$

Proof:
We write the expression (22) as

$$\frac{d[P_n(\tau)]}{d\tau} = \lambda_{n-1}(\tau;a)P_{n-1}(\tau) - \lambda_n(\tau;a)P_n(\tau) \qquad \text{for} \qquad n \geq 1.$$

By integration of the above expression with respect to τ between 0 and t, we get

$$\begin{aligned} \int_0^t d[P_n(\tau)] &= \int_0^t \lambda_{n-1}(\tau;a)P_{n-1}(\tau)d\tau - \int_0^t \lambda_n(\tau;a)P_n(\tau)d\tau \\ P_n(\tau)\big|_0^t &= Q_n(t) - Q_{n+1}(t) \qquad \text{for} \qquad n \geq 1. \end{aligned} \tag{36}$$

Since $P_n(0) = 0, \forall n \geq 1,$ so the proof is completed.

Corollary 1.7.1: Let $N(t)$ be an *NHP* with transition intensities given by (21), then

$$P[N(t) > n] = Q_{n+1}(t) \qquad \text{for} \qquad n \geq 0 \tag{37}$$

Proof: The proof consists of a direct calculation

$$\begin{aligned} P[N(t) > n] &= 1 - P[N(t) \leq n] \\ &= 1 - \sum_{j=0}^{n} P_j(t) = 1 - P_0(t) - \sum_{j=1}^{n} P_j(t) \end{aligned}$$

Using the previous result:

$$\begin{aligned} P[N(t) > n] &= 1 - P_0(t) - \sum_{j=1}^{n} \left[Q_j(t) - Q_{j+1}(t) \right] \\ &= 1 - P_0(t) - \left[Q_1(t) - Q_{n+1}(t) \right] \end{aligned} \tag{38}$$

Note that

$$Q_1(t) = \int_0^t \lambda_0(v; a) P_0(v) dv = -\int_0^t P_0'(v) dv = -P_0(v)|_0^t = 1 - P_0(t)$$

Replacing $Q_1(t)$ in (38) the proof is completed.
The expression (37) allows to calculate the *cdf* of an *NHP*.
Corollary 1.7.2: The function $Q_{n+1}(t)$ satisfies the following condition:

$$\lim_{t \to \infty} Q_{n+1}(t) = 1 \qquad \text{for} \qquad n \geq 0. \tag{39}$$

Proof:
From (37), we get

$$\lim_{t \to \infty} Q_{n+1}(t) = \lim_{t \to \infty} \left[1 - \sum_{j=0}^{n} P_j(t) \right].$$

As we have for $n \geq 1 : P_n(\infty) = 0$, and using the above relationship

$$\lim_{t \to \infty} Q_{n+1}(t) = 1 - \lim_{t \to \infty} P_0(t).$$

For example, from expression (9) when $a = 1$, we have:

$$P_0(t) = (1 + \kappa t)^{-\frac{q}{\kappa}} \qquad \text{for} \qquad \frac{q}{\kappa} > 0 \tag{40}$$

and we take the limit as $t \to \infty$, we get:

$$\lim_{t \to \infty} Q_{n+1}(t) = 1 - \lim_{t \to \infty} (1 + \kappa t)^{-\frac{q}{\kappa}} = 1. \square$$

Proposition 1.8: Let $N(t)$ be an *NHP* with transition intensities given by (21), then

$$\exp\left\{-\int_{t}^{t+h}\lambda_n(v;a)dv\right\}=\frac{P_0^{(n)}(t+h)}{P_0^{(n)}(t)}\qquad \text{for}\quad h\geq 0. \tag{41}$$

Proof:
By substituting (28) into (40), we have

$$\exp\left\{-\int_{t}^{t+h}\lambda_n(v;a)dv\right\}=\exp\left\{\int_{t}^{t+h}\frac{P_0^{(n+1)}(v)}{P_0^{(n)}(v)}dv\right\}$$

$$=\exp\left\{\int_{t}^{t+h}d\left[\ln\left(P_0^{(n)}(v)\right)\right]\right\}$$

$$=\exp\left\{.\ln\left[P_0^{(n)}(v)\right]\Big|_{t}^{t+h}\right\}=\frac{P_0^{(n)}(t+h)}{P_0^{(n)}(t)}.$$

Corollary 1.8.1: Let $N(t)$ be an NHP. If the probability that no event occurs in a small interval of length h is denoted by $P_0(t,t+h)$, that is $P_0(t,t+h)=P(N(t+h)-N(t)=0)$, then

$$P_0(t+h)=P_0(t)\cdot P_0(t,t+h)\qquad \text{for}\quad t,h\geq 0. \tag{42}$$

Proof:
According to Lundberg in [18]:

$$P(N(t+h)=0|N(t)=0)=\exp\left\{-\int_{t}^{t+h}\lambda_0(u)du\right\} \tag{43}$$

where $\lambda_0(t)$ denotes the intensity function associated with the time-dependent (or nonstationary) *PCP*. If we make $n=0$ in (40), then we obtain

$$P_0(t,t+h)=\exp\left\{-\int_{t}^{t+h}\lambda_0(v;a)dv\right\}=\frac{P_0(t+h)}{P_0(t)} \tag{44}$$

Thus,

$$P_0(t+h)=P_0(t)\cdot P_0(t,t+h)\qquad \text{for}\quad t,h\geq 0.$$

The expression obtained in (41) may be interpreted as if no event occurred, then the *NHP* has independent increments.

Lemma 1.9: Let $N(t)$ be an *NHP* with transition intensities given by (21). Then this CP satisfies

$$\sum_{j=0}^{m}\frac{\lambda_j'(t;a)}{\lambda_j(t;a)}=\lambda_0(t;a)-\lambda_{m+1}(t;a)\qquad \text{for all}\quad m\geq 0. \tag{45}$$

Proof:

From (25), we have

$$\frac{\lambda_j'(t;a)}{\lambda_j(t;a)} = \lambda_j(t;a) - \lambda_{j+1}(t;a) \qquad \text{for all} \quad j \geq 0. \tag{46}$$

Thus, (44) turns out the mth partial sum of a telescoping series and from here

$$\sum_{j=0}^{m} \frac{\lambda_j'(t;a)}{\lambda_j(t;a)} = \lambda_0(t;a) - \lambda_{m+1}(t;a) \qquad \text{for all} \quad m \geq 0.$$

Now, using the above lemma, we will prove the following proposition:

Proposition 1.10: Let $N(t)$ be an NHP with marginal pmf given by (5), then $P_n(t)$ satisfies that

i. Process with time-dependent increments

$$\lim_{h \to 0} \frac{P_{n,n+1}(t, t+h)}{h} = \lambda_n(t;a)$$

ii. The probability that no event occurs in $(t, t+h]$ is

$$P_0(t, t+h) = 1 - h\lambda_0(t;a) + o(h) \tag{47}$$

iii. The probability that one event occurs in $(t, t+h]$ is

$$P_1(t, t+h) = h\lambda_0(t;a) - o(h) \tag{48}$$

iv. *Faddy's conjecture*[2]: If the transition intensities be an increasing sequence with n, i.e,

$$\lambda_0(t;a) < \lambda_1(t;a) < \ldots < \lambda_n(t;a), \qquad \text{for any fixed } t \tag{49}$$

then $Var[N(t)] > \mathbb{E}[N(t)]$, this last inequality is reversed for a decreasing sequence.

Proof:

i. As the *NHP* is an *MPP* then, according to Lundberg in [18], for $0 \leq u < v, \quad i \leq j$, $N(t)$ satisfies:

$$\underbrace{P(N(v) = j \mid N(u) = i)}_{P_{i,j}(u,v)} = \binom{j}{i} \left(\frac{u}{v}\right)^i \left(1 - \frac{u}{v}\right)^{j-i} \frac{P_j(v)}{P_i(u)} \tag{50}$$

Replacing the expression $P_n(t)$ given in (12), when $\kappa \neq 0$, we obtain in (49) that the transition probabilities for the *NHP* are:

---

[2] See [21].

$$\begin{aligned}P_{i,j}(u,v) &= \binom{j}{i}\left(\frac{u}{v}\right)^i\left(1-\frac{u}{v}\right)^{j-i}\frac{P_j(v)}{P_i(u)}\\ &= \binom{j}{i}\left(\frac{u}{v}\right)^i\left(\frac{v-u}{v}\right)^{j-i}\left[\frac{\dfrac{(-1)^j v^j P_0^{(j)}(v)}{j!}}{\dfrac{(-1)^i u^i P_0^{(i)}(u)}{i!}}\right]\\ &= \frac{(u-v)^{j-i}}{(j-i)!}\frac{P_0^{(j)}(v)}{P_0^{(i)}(u)}\\ &= \prod_{m=1}^{j-i}\left[\frac{v-u}{m}\lambda_{m+i-1}(u;a)\right]\exp\left\{-\int_u^v \lambda_j(w;a)dw\right\}.\end{aligned} \tag{51}$$

We complete the proof of the theorem by the following steps: Rewrite the product in (50) by replacing all instances of $i = n, j = n+1, u = t$ and $v = t+h$, and we make the limit as h approaches zero. Then the transition intensities given by (21) represent the instantaneous transitions probabilities of the *NHP*.

ii. Certainly, the function given by (9) is continuous for $t \geq 0$ and also analytic, due to $P_0^{(n)}(t)$, exists for all $n \geq 1$. Then it is possible to express $P_0(t+h)$ through a Taylor series as follows:

$$P_0(t+h) = \sum_{m=0}^{\infty}\frac{h^m}{m!}P_0^{(m)}(t). \tag{52}$$

By substituting the expression for the mth derivative of $P_0(t)$ obtained given by (27) in (51), we have:

$$P_0(t+h) = P_0(t) + \sum_{m=1}^{\infty}\frac{h^m}{m!}\left[(-1)^m\left(\prod_{j=0}^{m-1}\lambda_j(t;a)\right)P_0(t)\right]. \tag{53}$$

Notice that $P_0(t+h)$ satisfies (41), then (52) is similar to:4

$$P_0(t)\cdot P_0(t,t+h) = P_0(t)\left[1+\sum_{m=1}^{\infty}(-1)^m\frac{h^m}{m!}\left(\prod_{j=0}^{m-1}\lambda_j(t;a)\right)\right] \tag{54}$$

Let $n = m-1$ then:

$$\begin{aligned}P_0(t,t+h) &= 1+\sum_{n=0}^{\infty}(-1)^{n+1}\frac{h^{n+1}}{(n+1)!}\left(\prod_{j=0}^{n}\lambda_j(t;a)\right)\\ &= 1-h\sum_{n=0}^{\infty}\frac{(-h)^n}{(n+1)!}\left(\prod_{j=0}^{n}\lambda_j(t;a)\right)\end{aligned} \tag{55}$$

From the expansion of the first terms of (54), we get:

$$P_0(t,t+h) = 1-h\lambda_0(t;a)+o(h) \tag{56}$$

where

$$o(h) = \sum_{n=1}^{\infty} \frac{(-h)^{n+1}}{(n+1)!} \prod_{j=0}^{n} \lambda_j(t; a).$$

The last function satisfies that $\lim_{h \to 0} o(h)/h = 0$ ([21, 22]).

iii. From (55) and the fact $P_0(t, t+h) = P(N(t+h) - N(t) = 0)$, we obtain

$$P(N(t+h) - N(t) > 0) = 1 - P_0(t, t+h). \tag{57}$$

Given that the *NHP* $N(t)$ is an *NHPBP* and assuming that we have in a small time interval, then there will be only two cases: there is a birth or not in that period. Thus,

$$P(N(t+h) - N(t) > 0) = P(N(t+h) - N(t) = 1) = P_1(t, t+h).$$

Then, from (56), we obtain:

$$P_1(t, t+h) = h\lambda_0(t; a) - o(h), \tag{58}$$

provided that h is infinitesimal.

iv. According to Steutel et al. in ref. [16], a non-degenerate distribution $\{P_n(t)\}$ is log-convex if and only if $P_n(t) > 0$ for all $n \geq 0$ and $\left\{\frac{P_{n+1}(t)}{P_n(t)}\right\}$ is a nondecreasing sequence. By assumption

$$\frac{P_n(t)}{P_{n-1}(t)} < \frac{P_{n+1}(t)}{P_n(t)} \qquad \text{for some} \quad n \geq 1 \tag{59}$$

By substituting (5) into (58)

$$\begin{aligned} \frac{\frac{t^n}{n!}\left[(-1)^n P_0^{(n)}(t)\right]}{\frac{t^{n-1}}{(n-1)!}\left[(-1)^{n-1} P_0^{(n-1)}(t)\right]} &< \frac{\frac{t^{n+1}}{(n+1)!}\left[(-1)^{n+1} P_0^{(n+1)}(t)\right]}{\frac{t^n}{n!}\left[(-1)^n P_0^{(n)}(t)\right]} \\ \frac{1}{n}\left(-\frac{P_0^{(n)}(t)}{P_0^{(n-1)}(t)}\right) &< \frac{1}{n+1}\left(-\frac{P_0^{(n+1)}(t)}{P_0^{(n)}(t)}\right) \\ \frac{1}{n}\lambda_{n-1}(t; a) &< \frac{1}{n+1}\lambda_n(t; a) \end{aligned},$$

we know $1 < \frac{n+1}{n}$ for all n. Hence, we have the following:

$$\lambda_{n-1}(t; a) < \frac{n+1}{n}\lambda_{n-1}(t; a) < \lambda_n(t; a). \tag{60}$$

Thus, we obtain that (48) is satisfied and, therefore, the conjecture holds.

The expression (48) allows to identify under- or over-dispersion of a *CP*, then we can classify the process according to the fixed criteria given in (16).

Corollary 1.10.1: If $a \neq 0$ and $N(t)$ is an *NHP*, then it does not have independent increments.

Proof:

From theorem 1.5, we know that an *NHP* is an *MPP*. According to McFadden in ref. [9], if $\{N(t), t \geq 0\}$ is a *CP* with independent increments, then its transition intensities satisfy that $\lambda_0(t; a) = \lambda_1(t; a)$, but by expression (48), we get

$$\lambda_0(t; a) = \frac{q}{(1+\kappa t)^a} \neq \frac{a\kappa}{1+\kappa t} + \frac{q}{(1+\kappa t)^a} = \lambda_1(t; a) \quad \text{if} \quad a \neq 0 \tag{61}$$

And therefore, $N(t)$ is a *CP* that does not have independent increments.

This was to be expected since that *MPP* has stationary increments but does not meet the condition of independent increments (see [23]).

6. Conclusions

In this chapter, we studied the *NHP* presenting some of its properties indicating that it is a good option for modelling *CP* regardless of the fact that it presents under- or over-dispersion.

Using transition intensities, we found some properties of the *NHP* and provided explicit analytic expressions for its *pmf* and *cdf*.

Author details

Gerson Yahir Palomino Velandia* and José Alfredo Jiménez Moscoso
Universidad Nacional de Colombia, Bogotá, Colombia

*Address all correspondence to: gypalominov@unal.edu.co

References

[1] Hofmann M. Über zusammengesetzte Poisson-Prozesse und ihre Anwendungen in der Unfallversicherung, PhD Thesis, Swiss Federal Institute of Technology in Zürich (ETH Zürich), Germany. 1955

[2] Walhin JF. Recursions for actuaries and applications in the field of reinsurance and bonus-malus systems, Université catholique de louvain, institut de statistique, Tesis Doctoral, Louvain-la-Neuve. 2000

[3] Thyrion P. Contribution a l'Etude du Bonus Pour Non Sinistre en Assurance Automobile. ASTIN Bulletin. 1960;**1**(3): 142-162

[4] Walhin JF, Paris J. Using mixed poisson processes in connection with Bonus-Malus systems. ASTIN Bulletin. 1999;**29**(1):81-99

[5] Walhin JF, Paris J. The Practical Replacement of a Bonus-Malus System. ASTIN Bulletin. 2001;**31**(2): 317-335

[6] Dubourdieu J. Remarques relatives à la théorie mathématique de l'assurance accidents. Bulletin Trimestriel de l' Institut des Actuaires Français. 1938;**44**: 79-126

[7] Grandell J. Mixed Poisson Processes, Chapman & Hall/CRC Monographs on Statistics & Applied Probability. New York: Chapman & Hall; 1997

[8] Karlis D, Xekalaki E. Mixed Poisson distributions. International Statistical Review. 2005;**73**(1):35-58

[9] McFadden JA. The Mixed Poisson process. Sankhyā: The Indian Journal of Statistics, Series A. 2002;**27**(1):83-92

[10] Jiménez M. Una relación entre la distribución de Hofmann y distribución de Panjer. Revista Integración. 2013; **31**(1):59-67

[11] Cox DR, Lewis PAW, The Statistical Analysis of Series of Events. Methuen's Monographs on Applied Probability and Statistics. London: Chapman and Hall; 1966

[12] Iseger PW, Dekker R, Smith MAJ. Computing compound distributions faster! Insurance: Mathematics and Economics. 1997;**20**:23-34

[13] Neyman J. On a New Class of "Contagious" distributions, applicable in entomology and bacteriology, institute of mathematical statistics. The Annals of Mathematical Statistics. 1939;**10**(1):35-57

[14] Katti SK, Gurland J. The Poisson Pascal Distribution. International Biometric Society. 1961;**17**(4):527-538

[15] Evans DA. Experimental evidence concerning contagious distributions in ecology. Biometrika. 1953;**40**(1): 186-211

[16] Steutel F, Van Harn K. Infinite Divisibility of Probability Distributions on the Real Line. New York: Marcel Dekker CRC Press; 2004

[17] Hansen BG. On log-concave and log-convex infinitely divisible sequences and densities. The Annals of Probability. 1988;**16**(4):1832-1839

[18] Lundberg OF. On Random Processes and Their Application to Sickness and Accident Statistics. 2nd ed. Almqvist & Wiksells; University of Virginia; 1964

[19] Seal HL. Stochastic theory of a risk business, Wiley Series in Applied

Probability and Statistics Series. New York: John Wiley & Sons, Inc.; 1969

[20] Walhin JF, Paris J. A general family of overdispersed probability laws. Belgian Actuarial Bulletin. 2002;**2**(1):1-8

[21] Faddy MJ. On variation in Poisson processes. Applied Probability Trust. 1994;**19**:47-51

[22] Serfling RJ. Approximation Theorems of Mathematical Statistics. New York: John Wiley & Sons; 1980

[23] Embrechts P, Frey R, Furrer H. Handbook of Statistics Vol. 19 "Stochastic Processes: Theory and Methods". North-Holland: Elsevier Science B.V; 2001

Chapter 5

Probability to Be Involved in a Road Accident: Transport User Socioeconomic Approach

Saúl Antonio Obregón Biosca, José Luis Reyes Araiza and Miguel Angel Pérez Lara y Hernández

Abstract

Road education is one of the most relevant issues focused to reduce traffic accidents, so it is important to analyze the driver's behavior on the roads. International research has found evidence for a relationship between socioeconomic characteristics and traffic accidents. In this sense, the chapter shows a methodology to estimate the probability to be involved in a road accident, considering the road education and the socioeconomic characteristics of the population of a specific region, taking the Santiago de Querétaro city (in México) as a study case. Through a logit model estimation and a survey applied to pedestrian, cyclist, motorcyclist, car driver, and freight driver allow us to determine which socioeconomic variables and road education are significant to determine the probability of being involved in a road accident.

Keywords: traffic accidents, probability, road education, socioeconomic level, transport modes, logit

1. Introduction

The present chapter shows one of the most relevant issues regarding the area of road safety since according to the WHO, road accidents are among the ten leading causes of death in the world [1]. "In Mexico, it is estimated that between 70% and 90% of traffic accidents are attributed to the driver, with human errors and driver offenses in traffic regulations as the two main contributing factors" [2].

Shell [3] exposes "Improving road education involves an analysis of human behavior, where both classroom instruction on safety issues, laws and regulations, vehicle operation, and those factors affecting driving are combined." It is for these reasons that "the vast majority of road education exams have focused on accidents" [4]. The factors in these studies include age, income, and driver's attitude.

In relation to the implementation of any road safety system, Ker *et al.* [5] and Mackay and Tiwari [6] acknowledge that human errors should be minimized in order to significantly improve road safety. In the circumstances of drivers, traffic safety policies recently implemented have been focused on improving their traffic behavior [7], particularly to endorse a better attitude when using roads [8–10]. Nonetheless,

Mirzaei *et al.* [7] reported that while many drivers show a positive safety attitude in regards to traffic, there are specific circumstances that may induce a poor traffic performance from some of these road users. Therefore, the authors inform us about the need to illuminate such situations, containing any potential cultural aspects. For the diverse groups of road users, Factor *et al.* [11, 12] proposed a theoretical model to analyze the influence that some social and cultural characteristics of these groups have on traffic safety, reporting that road safety differs in cultural and social features, including lifestyles and attitudes.

This study analyzed together the socioeconomic and road knowledge characteristics of these users to determine the probabilities of being involved in a road accident. This issue arises from the research that has been done [3], which informs us that those who have knowledge of road education are less likely to be involved in accidents or to carry out traffic violations. Whereas Factor *et al.* [12], using a logistic regression, found a relationship between socioeconomic status and presence in traffic accidents, as to say there is a direct correlation between higher level of education and greater socioeconomic status, which lowers the probability of being involved in a road accident.

It is worth mentioning that the present research aims to develop a methodology to create, step by step, a model that determines which socioeconomic variables and road education are significant to determine the probability of being involved in a road accident, which was applied to a case study in the city of Santiago de Querétaro.

This is why it is important to analyze the behavior of drivers on public roads since one of the main factors of road accidents is the lack of education and knowledge that these users may have about road safety. Not only does lack of knowledge influence road safety but also social factors, such as differing cultures, social behavior, the age of a driver, and the socioeconomic status of the drivers. This is an explanation as to the importance of doing the study because knowing these aspects that were previously mentioned, are all aspects that can attribute to a driver's performance when operating a vehicle.

2. Background

When examining the number of road mishaps as a meaning of a given country's economic level, Xu *et al.* [13] concluded that "road users' income is a determining factor for road safety." Concurring to its 2013 Global Status Report on road safety, such a conclusion is also reached by the World Health Organization, as low- and middle-income countries show higher traffic death rates when associated with high-income economies. Additional authors also report this cause–effect relationship [12, 14–16]. Overall, these authors claim that a low per capita income is a decisive factor for traffic crashes.

These accidents affect different social areas, and for this reason the subject of road education is a responsibility that belongs to a whole society, which encompasses pedestrians, cyclists, motorcyclists, drivers of vehicles, passengers, and transportation. Improving road education involves an analysis of human behavior, where both classroom instruction on safety issues, laws and regulations, vehicle operation, and those factors affecting driving, as well as vehicle driving practice are combined with a trained instructor [3]. It is for these reasons that the vast majority of road education exams have focused on accidents [4].

Regarding age, on the other hand, much of the road safety literature focuses on high-risk drivers, often being young, low-income men with low education [17]. It is recognized that older people appear to be more safety-conscious [18].

In terms of income, it should be noted that per capita income has been identified as a determinant of overall injury mortality [19]. Based on research conducted by Zmud and Arce [20, 21], it is ensured that lower-middle income groups may be at increased risk of occupant motor vehicle injuries. Attitude is a very important factor in road education, which also predicts longitudinally an unsafe driver [22].

2.1 Multi-criteria models for the decision-making process

For this process, three decision-making models are discussed that are based on the manipulation of the simple related data that provide the means to develop indicators in a systematic way [23]. These decision-making criteria represent a multi-criteria approach, which must be compared with other processes of several criteria such as the qualification model, the hierarchical analytical process (AHP), and the multiple attribute utility theory. The AHP method is a method that has been applied to deal with problems in different areas, matching the sentences of intangible qualitative criteria with tangible quantitative criteria [24]. The AHP method was initially developed by Saaty [25], with the objective of determining the relative importance of a set of alternatives in a multi-criteria decision problem. There are three main steps in the AHP: design of the hierarchy, a prioritization procedure, and the calculation of the results.

3. Methodology

Recent road safety research focuses on the need to improve the "behavior" of drivers [7]. In this sense, we did not give the task of evaluating 5 (five) road users, such as pedestrian, cyclist, motorcyclist, vehicle driver, and freight truck driver.

The study consisted of an evaluation of the previously mentioned users determined by a sample size as a significant representation; this evaluation was applied through a questionnaire designed for each type of user, which was divided into two parts; the first containing information such as general data, socioeconomic level, age and origin of acquired knowledge and accident, second is designed with information such as regulations or recommendations, traffic signals, current situation in road safety and human factors, infrastructure, courtesy and urbanity and applied situations. It should be noted that because each questionnaire was designed by user type there are variants in some questions.

This research also has an important message for society and aims to contribute knowledge on the subject as well as to help in the reduction of traffic accidents in our country. For the execution of this project, we will be using the five steps of methodology to conduct this investigation, we will also describe each of these steps:

3.1 Step 1: knowledge of the context of the variables to be evaluated and their development

The main objective of this stage consists of bounded problems for which the fundamental parameters can be defined. For this activity, some elements are incorporated in the analysis and are obtained from a review of global, national, and local literature in relation to safety education programs and driving tests. As a result of this analysis, a list of specific questions involving six common variables around which two or three user-related questions are written is based on the comparative analysis of the necessary knowledge. Each question was obtained through a review of the literature, the resulting number of questions for each of the users of the infrastructure is as follows: 24 for drivers of vehicles, 24 for freight conductors, 24 for motorcycle users, 21 for bicycle users, and 21 for pedestrians.

3.2 Step 2: structuring the questionnaire and evaluation

Within this stage, once the questions were established in the context of the selected variables two parallel processes will be carried out: the planning for the execution of the survey and the establishment of the weighting factors for the survey questions. The AHP method will be selected for this process, as it represents a structured and computerized process in which comparisons are made on a peer basis, which provides some evidence regarding the assessments made by experts of the Mexican Institute Transport (IMT) and the Autonomous University of Querétaro (UAQ). To obtain the reason scales of the AHP methodology, we compared the set of peer evaluations for each question. The peer comparison was as follows: 1 = equal, 3 = moderate, 5 = strong, 7 = very strong, and 9 = extreme.

3.3 Step 3: experimental design and sample size for survey operation

In this step, we will determine the size of the sample of users of the road examined which is calculated according to the number of inhabitants of the area [26] and the means of transport chosen by the users, as reported by Obregón and Betanzo [27].

$$n = \frac{N * Z\alpha^2 p * q}{d^2 * (N-1) + Z\alpha^2 * p * q} \tag{1}$$

Where N is the total number of inhabitants in the area (804 663 de Santiago de Queretaro), $Z\alpha$ = 1.96 (for a reliability 95%), p = expected proportion (in this case 5% = 0.05), q = 1 - p (in this case 1- 0,05 = 0,95), and d = precision (can be 1% to 3%; 2% was selected).

According to Eq. (1), 207 individuals were needed. This sample size considers individuals using the different means of transportation listed in **Table 1**, where it can be observed that freight vehicle, motorcycle, and bicycle users were the least frequent road users, with 1%, 1%, and 0.7%, respectively. To increase the reliability of these users, the sample size was increased to 20, for each of these modes. The number of validated questionnaires was 254.

The specific public areas for applying the survey were selected as a function of the type of transport infrastructure user: (1) public spaces, in which people spend at least 10 minutes completing some paperwork; (2) spaces around public schools, in which students move; and (3) recreational areas, in which users have more time to respond the survey (e.g. malls and public parks).

	Freight vehicle	Car	Motorcycle	Bicycle	Walk	Rest	Total
Distribution of users by mean of transport (%)	1.00	32.50	1.00	0.70	10.10	54.00	100
Percentage of sample size per type of infrastructure user	1	33	1	1	10	54	100
Estimated sampled	5	148	5	3	46	246	453
Total sampled	20	148	20	20	46	0	254

Table 1.
Sample and user distribution by transport means in Santiago de Querétaro. Own elaboration by the distribution data from Ref. [27].

3.4 Step 4: database processing

In this process, we will compile the database obtained through the questionnaires applied to each user evaluated. Subsequently, this database will be analyzed to know the socioeconomic and road users' knowledge. In the following graph (see **Figure 1**), the analysis of the variables of road education performed with the results obtained by the surveys in each one of the evaluated users is shown. This shows that the users that resulted with the lowest road knowledge in general are freight drivers (FD) and vehicle drivers (VD), unlike cyclists (C) who obtained the highest level of knowledge. At the same time, we can observe that the motorcyclists (M) obtained a low rating in regulation and recommendations (R&R); in contrast, the pedestrian (P) proved to have low knowledge in courtesy and urbanity (C&U).

The rest of the variables of road education by its initials are classified in the following form: traffic signals (TS), current situation in road safety and human factor (CRS&HF), infrastructure (Infra), and applied situations (AS).

3.5 Step 5: the probabilistic model

In the literature, the use of *Logit* models has been reported to estimate the probability of accidents [7, 28]. In this sense, the present research project estimated the presence of road accidents using *Logit* models. These models are estimated using the commercial software NLOGIT version 5, which was used for the same objective by Tay [29]; who mentions that binary regression models are adequate techniques to predict a binary dependent variable as a function of predictor variables.

Due to its ease in its estimation, the *logit* transformation is one of the most used in studies, this conducive search of a model of choice is more comfortable analytically, and the result was the binary *logit* model. This is under the assumption that ε_n is logistically distributed [29]; and the probability of choosing alternative i is given by Eq. (2).

$$P_n(i) = \frac{1}{1 + e^{-\mu(V_{in} - V_{jn})}} \tag{2}$$

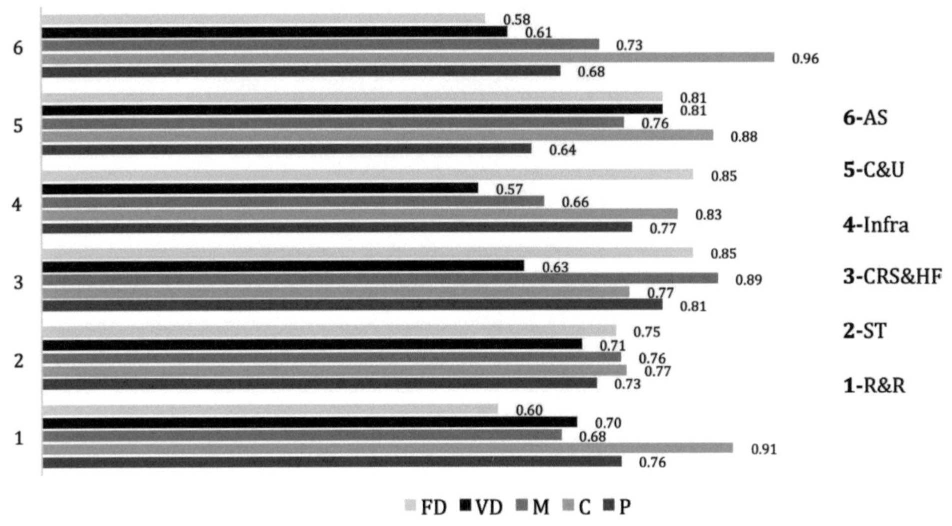

Figure 1.
Road education grade of each user.

For this model, the dependent variable $P(i)$, is a probability (between 0 and 1) that cannot be observed; only the choices of each individual are observed and these are variables (0 and 1).

4. Results and discussion

This section describes the *logit* models estimated to determine which socioeconomic and road education variables are significant to determine the probability of being involved in a road accident considering the means of transport used in their mobility. Depending on the mode of transport, the survey asks the user if they have been in a traffic accident in their life and during the last 12 months. Subsequently, each of the models obtained from each analyzed user is described. It should be noted that the first model (Model 1) was analyzed requesting the user if he has been involved in a traffic accident in his life. Unlike the second model (Model 2), which represents if you have been in a traffic accident in the last 12 months.

4.1 Freight driver

Two models were analyzed, in the first model, it can be seen that the significant variable is the income. Unlike Model 2, the most significant variable turned out to be the years with the driver's license (YDL) that the user has. It is worth mentioning that the variable that resulted most significantly in freight driver to determine the probability of being involved in a road accident is theincome (0.8345) (**Table 2**).

4.2 Vehicle driver

The first two models were analyzed, showing the following variables that are significant: if the user has a driver's license (DL) and the age at which the road

	Model 1		Model 2	
	Coef	SE Coef	Coef	SE Coef
Intercept	−3.7912*	2.0713	-21.4281 ·	16.6554
	(−1.830)		(−1.287)	
Income	0.8345*	0.4865	—	—
	(1.715)		—	
YDL	—	—	−0.7845 ·	0.6018
	—		(−1.304)	
*Note: ***, **, *, · = significance at 1, 5%, 10%, and 15% level.*				

Table 2.
Logit model, freight driver probability to be involved in a traffic crash.

	Model 1		Model 2	
	Coef	**SE Coef**	**Coef**	**SE Coef**
Intercept	−1.2401*	0.6994	−2.9609**	1.1795
	(−1.773)		(−2.510)	
Age	—	—	−0.8669***	0.2815
	—		−3.079	
Income	—	—	0.4208**	0.1791
	—		(2.349)	
DL	2.2472***	0.6498	2.4749**	1.0788
	(3.458)		(2.294)	
ARK	−0.4100**	0.1866	—	—
	(−2.197)		—	
*Note: ***, **, *, = significance at 1, 5%, and 10% level.*				

Table 3.
Logit models, vehicle driver probability to be involved in a traffic crash.

knowledge was obtained (ARK). In Model 2, the most significant variables were age (Age) and income (Income). It should be noted that the variable that resulted most significantly in vehicle drivers is driver's license (2.4749) (**Table 3**).

4.3 Motorcyclist

Two models were analyzed in Model 1, we can see the following variables that are significant: the level of road knowledge (LRK) and the courtesy and urbanity (C&U) that the user has. In Model 2, only a significant variable was obtained, which is the years with a driver's license (YDL) that the user has. It is worth mentioning that the variable that was most significant in motorcyclists to determine the probability of being involved in a road accident is the courtesy and urbanity (27.5462) (**Table 4**).

	Model 1		Model 2	
	Coef	**SE Coef**	**Coef**	**SE Coef**
Intercept	−5.9537 ·	3.7085	−4.8979 ·	3.6383
	(−1.605)		(−1.346)	
LRK	1.0178 ·	0.7227	—	—
	(1.408)		—	
YDL	—	—	−0.4753 ·	0.2785
	–		(−1.707)	
C&U	27.5462 ·	17.581	—	—
	(1.567)		—	

*Note: ***, **, *, · = significance at 1, 5%, 10%, and 15% level.*

Table 4.
Logit models, motorcyclist probability to be involved in a traffic crash.

	Model	
	Coef	**SE Coef**
Intercept	0.9418 ·	1.9526
	(0.482)	
Income	0.5037 ·	0.3304
	(1.524)	
C&U	−18.9062 ·	13.2425
	(−1.428)	

*Note: ***, **, *, · = significance at 1, 5%, 10%, and 15% level.*

Table 5.
Logit model, cyclist probability to be involved in a traffic crash.

4.4 Cyclists

For this user, only one model was analyzed, due to the fact that the data obtained show that they were not involved in an accident in the last 12 months. In the following model, the following variables were found to be significant: income (Income) and courtesy and urbanity (C&U) that these users may have on the infrastructure. It should be mentioned that the variable that was most significant in cyclists is the courtesy and urbanity of users (−18.9062) (**Table 5**).

4.5 Pedestrian

As we analyzed Model 1, we can see the following variables that were significant: age (Age) and applied situations of users (AS). In contrast to Model 2, the most significant variables were the income (Income), level of road knowledge (LRK) they believe they have, and the age at which they obtained road knowledge (ARK). The significant variables that influence the probability of the pedestrian being involved in a traffic accident are applied situations (−10.2266) and the age at which they obtained road knowledge (−1.2199) (**Table 6**).

	Model 1		Model 2	
	Coef	**SE Coef**	**Coef**	**SE Coef**
Intercept	−1.286*	0.6501	−6.0174*	3.2642
	(−1.859)		(−1.843)	
Income	—	—	0.8987*	0.4893
	—		(−1.837)	
Age	0.5841**	0.2457	—	—
	(−2.377)		—	
LRK	—	—	1.139*	0.6655
	—		(−1.712)	
ARK	—	—	−1.2199*	0.685
	—		(−1.781)	
AS	−10.2266*	5.358	—	—
	(−1.909)		—	

*Note: ***, **, *, · = significance at 1, 5%, 10%, and 15% level.*

Table 6.
Logit model, pedestrian probability to be involved in a traffic crash.

5. Conclusions

The chapter shows how the statistical *logit* probability model can characterize the effect of socioeconomic and educational factors on the population and the probability of being involved in a traffic accident. The overall result for the population surveyed identify both the level of road education and the income of the users' infrastructure. The significant variables that influence the probability of the user being involved in a traffic accident by transport mode are as follows:

Amongst freight drivers, it was found that the most significant variables influencing the probability of being involved in a road accident are income and years with a driver's license. Vehicle drivers, age (Age), income (Income), if you have a driver's license (DL), and the age at which you gained road knowledge (ARK) were found to be the most significant variables to determine the probability of being in a road accident. It was found that for motorcyclists the factors were the level of road knowledge (LRK) they were considered to have, years of driver's license (YDL) and Courtesy and Urbanity (C&U) as being the most significant variables for these users. For cyclists, it was found that income as well as courtesy and urbanity were the most significant variables. On the other hand, for pedestrians, it was found that the income, age, level of roadway knowledge that they considered to have, the age at which they obtained road knowledge, and the situations applied were the most significant variables.

In the case of motorized means of transport, the following aspects should be considered; age of users, socioeconomic characteristics, age and origin of acquired knowledge, and courtesy and urbanity. In the case of nonmotorized means of transport, the aspects to be taken into account are age, socioeconomic characteristics, age and origin of acquired knowledge, courtesy and urbanity, and the situations applied in this way.

The results of this research can be useful in defining road safety policies. In this sense, Mirzaei et al. [7] suggest that campaigns could be carried out to strengthen educational programs to minimize the probability of road accidents, considering the socioeconomic status and road education aspects of road users.

Conflict of interest

The authors declare no conflict of interest.

Author details

Saúl Antonio Obregón Biosca*, José Luis Reyes Araiza
and Miguel Angel Pérez Lara y Hernández
Faculty of Engineering, Autonomous University of Querétaro, Santiago de Querétaro, México

*Address all correspondence to: saul.obregon@uaq.mx

References

[1] World Health Organization (WHO). Global Status Report on Road Safety 2018. Geneva: World Health Organization; 2018. p. 420

[2] García J, Acosta S, Vázquez C. Educación vial y sustentabilidad: Hacia una convivencia y equilibrio urbano México. Mexico: Universidad Autónoma del Estado de México; 2010. p. 152

[3] Shell DF, Newman IM, Córdova-Cazar AL, Heese JM. Driver education and teen crashes and traffic violations in the first two years of driving in a graduated licensing system. Accident; Analysis and Prevention. 2015;**2015**(82):45-52. DOI: 10.1016/j.aap.2015.05.011

[4] Lonero L, Mayhew D. Large-scale Evaluation of Driver Education: Review of the Literature on Driver Education Evaluation 2010 Update. Washington, D.C.: AAA Foundation for Traffic Safety; 2010. DOI: 10.1037/e612252011-001

[5] Ker K, Roberts I, Collier T, Beyer F, Bunn F, Frost C. Post-licence driver education for the prevention of road traffic crashes: A systematic review of randomised controlled trials. Accident; Analysis and Prevention. 2005;**37**:305-313. DOI: 10.1016/j.aap.2004.09.004

[6] Mackay M, Tiwari G. The prevention of road traffic injuries. In: Proceedings of the WHO Conference of Road Traffic Injury Prevention. Geneva, Switzerland: WHO; 2001. pp. 24-32

[7] Mirzaei R, Hafezi-Nejad N, Sabagh MS, Moghaddam AA, Eslami V, Rakhshani F, et al. Dominant role of drivers' attitude in prevention of road trafficcrashes: A study on knowledge, attitude, and practice of drivers in Iran. Accident; Analysis and Prevention. 2014;**66**:36-42. DOI: 10.1016/j.aap.2014.01.013

[8] Wang Y, Zhang W, Reimer B, Lavalliere M, Lesch MF, Horrey WJ, et al. The effect of feedback on attitudes toward cellular phone use while driving: A comparison between novice and experienced drivers. Traffic Injury Prevention. 2010;**11**(5):471-477. DOI: 10.1080/15389588.2010.495761

[9] Martinov-Cvejin M, Jakovljevic D, Nalcic B, Grujic V, Ac-Nikolic E. Knowledge, attitude and practice in school children regarding traffic accident injuries. Medicinski Pregled. 1993;**46**(9-10):349-352

[10] Teoh YL, Soh E, Heng J, Heng BH, Cheah J. A KAP survey of evidence-based medicine and clinical practice guidelines among primary care doctors in Singapore. Annals, Academy of Medicine, Singapore. 2004;**33**(5 Suppl.):S32-S33

[11] Factor R, Mahalel D, Yair G. The social accident: A theoretical model and a research agenda for studying the influence of social and cultural characteristics on motor vehicle accidents. Accident; Analysis and Prevention. 2007;**39**:914-921. DOI: 10.1016/j.aap.2006.12.015

[12] Factor R, Mahalel D, Yair G. Inter-group differences in road- traffic crash involvement. Accident; Analysis and Prevention. 2008;**40**:2000-2007. DOI: 10.1016/j.aap.2008.08.022

[13] Xu J, Kockelman KM, Wang Y. Modeling crash and fatality counts along mainlanes and frontage roads across Texas: The roles of design, the built

environment, and weather. In: 93rd Transportation Research Board Annual Meeting. Washington: Transportation Research Board; 2014

[14] Shinar D. Demographic and socioeconomic correlates of safety belt use. Accident; Analysis and Prevention. 1993;**25**:745-755. DOI: 10.1016/0001-4575(93)90038-x

[15] Braver ER. Race, Hispanic origin, and socioeconomic status in relation to motor vehicle occupant death rates and risk factors among adults. Accident; Analysis and Prevention. 2003;**35**:295-309. DOI: 10.1016/s0001-4575(01)00106-3

[16] Zambon F, Hasselberg M. Socioeconomic differences and motorcycle injuries: Age at risk and injury severity among young drivers—a Swedish nationwide cohort study. Accident; Analysis and Prevention. 2006;**38**:1183-1189. DOI: 10.1016/j.aap.2006.05.005

[17] Shinar D, Schechtman E, Compton RP. Trends in safe driving behaviors and in relation to trends in health maintenance behaviors in the USA: 1985-1995. Accident; Analysis and Prevention. 1999;**31**:497-504. DOI: 10.1016/s0001-4575(99)00006-8

[18] Boyle J, Dienstfrey S, Sothoron A. National Survey of Speeding and Other Unsafe Driving Actions. Washington, DC: US Department of Transportation; 1998. p. 89

[19] Baker SP, O'Neill B, Ginsburg MJ, Li G. The Injury Fact Book. 2nd ed. New York: Oxford University Press; 1992. p. 368

[20] Zmud JP, Arce CH. The influence of consumer culture and race on travel behavior. Personal Travel: The Long and Short of it. In: Conference Proceedings Transportation Research Board. Washington: Federal Highway Administration; 1999

[21] US Census Bureau. Educational Attainment in the United States, 1995. Washington, DC: US Department of Commerce; 1995. p. 103

[22] Iversen H, Rundmo T. Attitudes towards traffic safety, driving behaviourand accident involvement among the Norwegian public. Ergonomics. 2004;**47**(5):555-572. DOI: 10.1080/00140130410001658709

[23] Oswald M, McNeil S. Rating sustainability: Transportation investments in urban corridors as a case study. Journal of Urban Planning and Development. 2010;**136**(3):177-185. DOI: 10.1061/(ASCE)UP.1943-5444.0000016

[24] Betanzo E, Obregón S, Romero J. Testing a new methodology to assess urban freight systems through the analytic hierarchy process. Modern Traffic and Transportation Engineering Research. 2013;**2**(2):78-86

[25] Saaty TL. Decision Making for Leaders: The Analytical Hierarchy Process for Decision in a Complex World. Belmont: Lifetime Learning Publications; 1982. p. 292

[26] INEGI. Censo de población y vivienda 2010. México: INEGI; 2010

[27] Obregón S, Betanzo E. Los viajes urbanos en una ciudad media mexicana, caso de estudio: Santiago de Querétaro Economía. Sociedad y Territorio. 2015;**XV**(47):61-98. DOI: 10.22136/est002015554

[28] Shinar D, Schechtman E, Compton R. Self-reports of safe driving behaviors in relationship to sex, age,

education and income in the US adult driving population. Accident; Analysis and Prevention. 2001;**33**:111-116. DOI: 10.1016/s0001-4575(00)00021-x

[29] Tay R. A random parameters probit model of urban and rural intersection crashes. Accident; Analysis and Prevention. 2015;**84**:38-40. DOI: 10.1016/j.aap.2015.07.013

Chapter 6

Quantifying Risk Using Loss Distributions

Retsebile Maphalla, Moroke Mokhoabane, Mulalo Ndou and Sandile Shongwe

Abstract

Risk is unavoidable, so quantification of risk in any institution is of great importance as it allows the management of an institution to make informed decisions. Lack of risk awareness can lead to the collapse of an institution; hence, our aim in this chapter is to cover some of the ways used to quantify risk. There are several types of risks; however, in this chapter, we focus mainly on quantification of operational risk using parametric loss distributions. The main objective of this chapter is to outline how operational risk is quantified using statistical distributions. We illustrate the application of parametric loss distributions' risk quantification using "Taxi claims data" which seems to best fit one of the loss distributions and fully illustrate how to quantify this specific data. More importantly, we also illustrate how to implement quantification of risk for two other scenarios: (i) if we assumed the underlying distribution is unknown and use the nonparametric empirical distribution approach, and (ii) when using the generalized extreme value (GEV) distribution approach. The latter two scenarios were not the main objective but were done in an effort to compare our results with some of the more commonly used techniques in real-world risk analysis scenarios.

Keywords: risk quantification, loss distributions, parametric, nonparametric, value-at-risk

1. Introduction

The topic presented in this chapter serves to give novice risk readers an idea of how institutions quantify their operational risks using parametric loss distributions. For various financial institutions, risk is classified into different components. Firstly though, risk is defined as the probability of an event and the potential loss. Put differently, [1] defined risk as a condition in which there is a possibility of an adverse deviation from the desired outcome that is expected or hoped for. Secondly, [2] provides an excellent account of four main categories of financial risks (more applicable in the banking sector), i.e. credit risk, market risk, operational risk, and others. For more discussion on some of the latter mentioned risks, see [3]'s Chapter 7 and the corresponding tools and techniques discussed in [3]'s Chapter 8. In this chapter, we are focusing mainly on the quantification of operational risk.

There have been multiple instances in the previous century of many big multinational firms experiencing total collapse due to a lack of risk control. For instance, employees may embezzle funds from the firm, rogue employees may make unauthorized deals, etc. For best examples of the latter, see Chapter 1 of [2] and Chapter 20 of [3]. Note though, in the South African context, the best example for poor operational risk management are Steinhoff, Hullett, Venda Building Society (VBS) mutual bank, Eskom, South African Airways (SAA) and more recently the Capitec bank computer systems failure during a peak period of the month in 2022. It is worth mentioning that a variety of sources have indicated that, in most instances, losses incurred due to operational risk normally would originate from poor management practices, outsourcing nonstrategic activities, or external factors.

In this chapter, a study on operational loss data will be conducted. We hope to determine the loss distribution that best fits the data by performing goodness-of-fit tests to the proposed models and estimating the parameters using appropriate statistical methods so that it can be possible to forecast or quantify the loss to be anticipated.

To date, financial institutions are making it a norm to manage their exposure to different types of risks, see [4]. Quantification of risk is of great importance, a proper evaluation of risk in any financial institution is an uncertainty problem that may easily lead to the bankruptcy of that firm and would consequently become a major concern for national and international financial regulatory bodies. This research work is compiled to contribute to the improvement of the quantification of operational risk using the loss distribution approach (LDA). According to [5], operational risk is the probability of loss resulting from insufficient or unsuccessful internal processes, people, and systems or from external events. Consequently, in the next section, we review the five most common parametric loss distributions namely: Pareto, Burr, gamma, Weibull, and log-normal distributions. These loss distributions are reviewed mainly in the aspect of quantification of operational risk.

This topic is applicable to a wide variety of fields as all institutions face some certain type of risk which if left unnoticed and unmanaged, could lead to total collapse of the firm or the worldwide economy (as seen in the last two global financial crises—the domino effect). Operational risk is quantified in several institutions; according to [6], this is done because we cannot predict the future for certain, but we can prepare and anticipate it. Risk quantification gives us an insight into what we can anticipate. Quantification of risk is done in several financial institutions, e.g. banks, universities, insurance companies, etc. The limitation of our research is as follows: it is applicable in scenarios when the underlying operational loss data fits (or almost fits) the loss distributions considered here (i.e. Pareto, Burr, gamma, Weibull, and log-normal distributions). In the event of the data not passing the goodness-of-fit tests for any of the latter distributions, then in the concluding section (i.e. Section 4), we shall list different alternatives approaches that the readers need to consider.

Note that the field of risk identification and quantification has become more important as globalization is expanding. To date, different financial institutions are realizing the importance of quantifying risk to avoid huge losses that may even result in bankruptcy. The aspect of risk quantification is pivotal in making the best business decisions.

Therefore, the rest of the chapter is structured as follows: in Section 2, we review several publications that have covered operational risk using different loss distributions. Moreover, we take note of various approaches that were used to quantify risk exposure. Next, in Section 3, we use a dataset to illustrate quantification of risk using loss distributions. Given that this research work is a continuation of previous literature studies, in Section 4, we provide some concluding remarks and offer several possible future research topics.

2. Literature review

2.1 Introduction

The three major classes of financial risks and their corresponding definitions are defined as follows:

- Market risk is the risk inherent from exposure to capital markets; see [3]. For instance, if some event led to insurance companies paying out large claims or banks are being exposed to risk due to an adverse movement in the stock market.
- Credit risk arises when losses are observed due to the inability of a debtor to perform an obligation in accordance with agreed terms; see [2].
- Operational risk is the risk of loss resulting from inadequate or failed internal processes, people, or systems or external events; see [7].

Note that [8] argued that the probability of an operational risk event increases with many personnel and with a greater transaction volume. The latter is also based on the study by [9] who investigated the effect of bank size on operational loss amounts and deducted that, on average, for every unit increase in bank size, operational losses are predicted to increase by approximately a fourth of a root of that. Note that there are different classes of operational losses that the financial industry must be aware of; see [2]:

i. high frequency and high magnitude

ii. high frequency and low magnitude

iii. low frequency and low magnitude

iv. low frequency and high magnitude.

Category (i) has been argued that it is not feasible/implausible in the financial industry, with (ii) and (iii) are unimportant and can often be both prevented. However, category (iv) tends to cause the most devastation loses, with the best example being the 1995 Barings Bank's collapse (also portrayed in the movie "Rogue Trader"). Consequently, banks must be extremely cautious of these types of losses as they tend to cause bankruptcy in many financial institutions. Low-frequency/high-severity operational losses can be extreme in size when they are compared to the rest of the data. If you construct a histogram of the loss distribution, the low-frequency/high-severity operational losses events would be placed in the far-right end, which often referred to as "tail event". Due to operational loss, data exhibit such tail events. We say that the data are heavy-tailed.

In different fields that use "Data Science" techniques (e.g. insurance, banks, etc.), different types of distributions are used to model data due to the different products that are offered by several financial institutions. These financial institutions are increasingly measuring and managing their exposure to different types of risks; see [4]. A proper evaluation of risk in any financial institution is an uncertainty problem that may easily lead to a bankruptcy of that firm and consequently is a major concern for national and international financial regulatory bodies.

It is important to mention that risk data from different products offered by financial institutions (e.g. micro-insurance, re-insurance, investment, savings, stock exchange, etc.) are distributed differently; see for instance [10]. Consequently, a thorough understanding of a variety of distributions is a must for an inspiring data scientist. According to [2], the most applied basic distributions in quantifying operational risk are those that are skewed to the right (right-tailed). There are two main ways to categorize the right-tailed loss distributions, i.e. parametric and nonparametric approaches. More specifically, in this chapter, we will consider the following most common parametric loss distributions: (i) exponential, (ii) gamma, (iii) Weibull, (iv) Pareto, (v) Burr, and (vi) log-normal. It is worth mentioning that these are not the only existing loss distributions, for example, a combination of two of the above, i.e. the composite Weibull-Pareto distribution in the context of risk is discussed in [11].

The next subsections discuss the following: Section 2.2 provides some distributional properties of the considered parametric loss distributions. More importantly, Section 2.2 provides the literature review of some publications that applied the considered distributions in the context of risk analysis. Next, Section 2.3 gives a brief discussion on nonparametric loss distributions (seldomly used), and Section 2.4 discusses some well-known methods of quantifying risk. Section 2.5 discusses other types of risks that use the LDA. Finally, Section 2.6 gives some concluding remarks.

2.2 Parametric loss distributions

A summary of some publications that discussed parametric loss distribution's application in operational risk is provided in **Table 1**. This table was constructed with an effort to easily identify which type of loss distributions is discussed in these separate publications. The corresponding loss distribution function properties are listed in **Table 2** with the expressions adopted from [12, 13] and Chapter 6 of [2].

Note that when the different parameters in **Table 2** are varied, the distributions tend to vary significantly, especially in the tail area. The latter will be illustrated in detail in the next section.

2.2.1 Pareto distribution

The Pareto distribution is a very heavy-tailed distribution that takes on positive values, and its parameter α is used to determine the size of the tail heaviness. The Pareto distribution tail is monotonically decreasing, and this means that the tail decreases as x increases and it becomes is thicker for values of x closer to zero. To derive the Pareto distribution, assume that a variate x follows an exponential distribution with mean β^{-1}; furthermore, suppose that β follows a gamma distribution, therefore the x follows a Pareto distribution; see [17]. Note that when $\alpha < 1$, a very heavy tail is encountered with the mean and variance being infinite. This means that losses of infinite sizes are theoretically possible. The extreme heaviness of the Pareto distribution tail makes it ideal for modeling losses of high magnitudes; see [2]. There are also different versions of the Pareto distribution that are used in risk analysis. The most popular of those variations is being the generalized Pareto distribution (GPD).

Publication	Loss distribution					
	Gamma	*Pareto*	*Weibull*	*Log-normal*	*Burr*	*Other*
[2]	✓	✓	✓	✓	✓	
[4]	✓					
[5]				✓		
[10]		✓				
[11]		✓	✓			✓
[12]	✓					✓
[13]			✓			✓
[14]	✓		✓			
[15]					✓	✓
[16]	✓	✓	✓	✓	✓	
[17]	✓	✓	✓	✓	✓	✓
[18]				✓		✓
[19]					✓	
[20]					✓	
[21]					✓	
[22]	✓					

Table 1.
A summary of publications discussed in this chapter and their classification according to the type of loss distribution.

The GPD is especially good while modeling data greater than a high threshold, also known as estimation of tails of extreme losses.

2.2.2 Burr distribution

The Burr distribution is heavy-tailed, and it is skewed to the right; see [19]. The Burr distribution is a special case of GPD described in subsection 2.2.1. It has three parameters which gives it more flexibility over the traditional Pareto distribution. The Burr distribution has an additional parameter γ and when $\gamma = 1$, it reduces to a Pareto distribution. One of the well-known uses of the Burr distribution is modeling natural catastrophes and as a result, it is a popular distribution or model for use in the insurance industry for pricing of premiums; see [2, 22]. The family of Burr distributions goes back to 1941, and it is sometimes referred as the extended Pareto or beta prime distribution. All the PDFs of the loss distributions in the Burr family have a monotonically decreasing, right-skewed tails; see [15]. The Burr distribution is well recognized in probability theory with many applications in agriculture, biology, etc., see [20].

2.2.3 Gamma distribution

The gamma distribution is a light-tailed distribution which is skewed to the right. The gamma distribution is a two-parameter distribution that is a generalization of the exponential distribution. It is a two-parameter probability distribution, where x is a

Loss distribution	PDF - $f(x)$	CDF - $F(x)$	Domain of variable/parameter (s)	Mean	Variance
Exponential	$\lambda e^{-\lambda x}$	$1-e^{-\lambda x}$	$x>0,\lambda>0$	$\frac{1}{\lambda}$	$\frac{1}{\lambda^2}$
Gamma	$\frac{\beta^\alpha}{\Gamma(\alpha)}(x)^{\alpha-1}e^{-\beta x}$	$\Gamma(\alpha,\beta x)$	$x>0,\alpha>0,\beta>0$	$\frac{\alpha}{\beta}$	$\frac{\alpha}{\beta^2}$
Weibull	$\alpha\beta(x)^{\alpha-1}e^{-\beta x^\alpha}$	$1-e^{-\beta x^\alpha}$	$x>0,\beta>0,\alpha>0$	$\beta^{-\frac{1}{\alpha}}\Gamma(1+\frac{1}{\alpha})$	$\beta^{-\frac{1}{\alpha}}\Gamma(1+\frac{1}{\alpha})$
Pareto	$\frac{\alpha\beta^\alpha}{x^{\alpha+1}}$	$1-\left(\frac{\beta}{x}\right)^x$	$\alpha>0,\beta<x<\infty,\beta>0$	$\frac{\alpha\beta}{\alpha-1}$	$\frac{\alpha\beta^2}{(\alpha-1)^2(\alpha-2)}$
Burr	$\gamma\beta^\alpha\frac{x^{\gamma-1}}{(\beta+x^\gamma)^{\alpha+1}}$	$1-\left(\frac{\beta}{\beta+x\gamma}\right)^\alpha$	$x>0,\gamma>0,\alpha>0,\beta>0$	$\frac{\beta^{\frac{1}{\gamma}}}{\Gamma(\alpha)}\Gamma\left(1+\frac{1}{\gamma}\right)\Gamma\left(\alpha-\frac{1}{\gamma}\right)$	$\frac{\beta^{\frac{2}{\gamma}}}{\Gamma(\alpha)}\Gamma\left(1+\frac{2}{\gamma}\right)\Gamma\left(\alpha-\frac{2}{\gamma}\right)-\frac{\beta^{\frac{2}{\gamma}}}{\Gamma^2(\alpha)}\Gamma^2\left(1+\frac{1}{\gamma}\right)\Gamma\left(\alpha-\frac{1}{\gamma}\right)$
Log-normal	$\frac{1}{\sqrt{2\pi}\sigma x}e^{\frac{-(\log x-u)^2}{2\sigma^2}}$	$\Phi\left(\frac{\log x-\mu}{\sigma}\right)$	$x>0,-\infty<\mu<\infty,\sigma>0$	$e^{\mu+\frac{\sigma^2}{2}}$	$\left(e^{\sigma^2}-1\right)e^{2\mu+\sigma^2}$
where $\Gamma(\alpha,\beta x)\sqrt{\frac{2\pi}{\alpha}}\left[\frac{1}{e}\left(\alpha+\frac{10\alpha}{120\alpha^2-1}\right)\right]^\alpha-\frac{(\beta x)^\alpha}{\alpha e\beta^{\beta x}}\sum_{n=0}^{15}\frac{(\beta x)^n(\alpha)!}{(\alpha+n)!}$					

Table 2.
Some properties of different loss distributions.

random variable, β is a scale parameter, α is the shape parameter, and $\Gamma(\bullet)$ is a gamma distribution; see **Table 2**. The gamma distribution is said to be the generalization of the exponential distribution because for $\gamma = 1$ it becomes the exponential with parameter $\lambda = 1/\beta$, and it is usually used to model time between events; see [2]. According to [7], the gamma distribution is one of the most important loss distributions in risk analysis because it forms the base for creating many of the popular distributions we have. Exponential distribution is described by a density f and distribution F given in **Table 2** above, where λ represents the "failure" rate. The distribution is tractable and has unique mathematical properties, e.g. the failure distribution is described by a single parameter known as the mean time to failure, denoted by θ, also that the failure rate is defined by knowing the mean life, i.e. $\lambda = 1/\theta$. The exponential distribution can be used to model the time elapsed until the next event (e.g. accident); see [23].

The exponential PDF has a monotone decrease and an exponentially decreasing and light tail. This means that when it is applied in risk analysis, the event of high losses is given an almost zero probability; see [2]. Due to this property, [14] stated that the exponential distribution is not used very much in operational losses, but the constant decrease of the tail is useful for modeling lifetime data of items which have a constant failure rate. The exponential distribution has attractive and easily understandable mathematical properties; thus, it is mostly used in risk analysis for developing other models.

2.2.4 Weibull distribution

Another generalization of the exponential distribution is the Weibull distribution, and it has two parameters (see **Table 2**) compared to the one parameter of the exponential distribution. The Weibull distribution has a light tail which is skewed to the right. The additional parameter allows the Weibull distribution to have more flexibility as well as heavier or lighter tail than the exponential distribution. That is, the Weibull distribution has a lighter tail than exponential distribution if $\alpha < 1$, equals to the exponential distribution if $\alpha = 1$ and has a heavier tail than exponential distribution if $\alpha > 1$; see [2]. Furthermore, [2] stated that in risk analysis, the heavy-tailed Weibull distribution is a popular model as it has been shown to be optimal for modeling asset returns as well as used in reinsurance.

2.2.5 Log-normal distribution

A log-normal distribution is a moderately heavy-tailed distribution that is skewed to the right. The distribution is derived by taking the natural logarithm of the data and fitting it to the normal distribution. The distribution is right-tailed and takes on only positive x values. The log-normal distribution, like the normal distribution, has parameters μ and σ (see **Table 2**). The distribution is useful for modeling of claim sizes. The thick tail and right skewness properties make it fit many situations. The log-normal can also resemble the normal distribution if the α is very small, and this property is not always desirable for analyzing risk; see [16].

2.3 Nonparametric loss distribution

In the nonparametric loss distributions (e.g. empirical distribution function), all the data on the certain risk type is considered. In other words, we do not have to estimate any parameters as all the data are available (which is hardly ever the case); see [2, 16]. According to [7], the CDF of the empirical distribution, $F_n(x)$, is given by:

$$F_n(x) = \frac{1}{n}\#\{i : x_i \leq x\} \tag{1}$$

where # denotes the number of observations $\leq x$, and n is the total number observations in the sample.

Some of the advantages of nonparametric loss distributions are as follows:

- It does not assume any underlying distribution, thus letting the data speak for itself.
- It is simple and easy to understand and does not require any complicated sample theory.
- It might be the only alternative for small sample sizes.

Below are some of the disadvantages:

- Less efficient to compute and may provide inaccurate results, especially when the underlying distribution is known.

2.4 Risk quantification

According to [18, 24], LDA is widely used to quantify operational risk; moreover, both [18, 24] showed that when quantifying operational risk, the PDF for an occurrence and the frequency for that occurrence are approximated firstly for a certain risk type or business line then later for the institution. The process of deriving these probability distributions is done in three steps: firstly, the loss severity distribution is derived; secondly, the loss frequency distribution is also derived; and lastly, the aggregate loss distribution is found by compounding the severity and frequency loss distributions.

The Value-at-Risk (VaR), which is a combination of expected and unexpected losses, is used when approximating the PDFs, and [18, 24] stated that the Capital-at-Risk (CaR) given in Eq. (2) is just the VaR, and this value is computed for a certain risk type cell and a certain occurrence type:

$$CaR\ (i, j; \alpha) = EL\ (i, j) + UL\ (i, j; \alpha) \tag{2}$$

Note that in Eq. (2), we use the indices i and j to denote a given business line and a given event type, $EL(i, j)$ is the expected loss, and $UL(i, j; \alpha)$ is the unexpected loss at significance level α.

Another method of quantifying risk is the internal measurement approach (IMA). According to [18], when using IMA, the business type and the event type risk are both quantified using

$$CaR\ (i, j) = EL\ (i, j) \times \gamma\ (i, j) \times RPI\ (i, j) \tag{3}$$

where γ is the scaling factor and RPI is the risk profile index.

LDA is of great importance when computing regulatory capital, and as noted by [18], even though LDA is such a great tool, it also has its downside. This is due to a lack of data and even if a bank keeps large amounts of losses data, they may still be unrepresentative of potential extreme losses. Three of some popular approaches under LDA is the extreme value theory (EVT), VaR, and IMA which are briefly discussed in Section 3.

Risks need to be measurable so that they can be evaluated and examined. It is ideal to have a high-quality historical data that can be subjected to in-depth statistical analysis. Numerous quantification models tend to lack high-quality statistical data. The types of risk determine which quantification technique to use. In finance, the main methods to quantify risk are as outlined below; see [2, 3]:

- Dynamic financial analysis

This simulates the enterprise's overall risks as well as their interactions. Typically, forecast balance sheets and projected income statements are produced as outputs using cashflows.

- Financial Conditions Reports (FCR)

The Financial Conditions Report (FCR) displays both the current state of solvency and potential future developments. The volume and profitability of new business as well as any special characteristics it might have would typically be projected.

- Quantitative methods

Quantitative methods are employed for risks in insurance and underwriting, markets, and economies, such as interest rate, basis risk, and market fluctuations. Time series and scenario analysis might be included, as well as the fitting of statistical models and subsequent calculation of risk metrics like VaR.

- Credit risk models

Instead of measuring the risk in a credit portfolio, these models assess the credit risk of a single entity (business or person). These may be quantified as well as subjectively, and counterparty risk is one of them. A credit risk model's job is to take the state of the overall economy and the circumstances surrounding the company under consideration as inputs and provide a credit spread as an output. In this context, structural and reduced form models are the two main groups of credit risk models. Based on the value of a company's assets and obligations, structural models are used to assess the likelihood that a default will occur.

- Asset Liability Modeling (ALM)

This approach, which is common in the insurance industry and primarily measures liquidity and capital requirements, might be used by various types of financial companies.

- Scenario analysis

Operational hazards and other risks that are challenging to measure, such as legal risk, regulatory risk, agency risk, moral hazard, strategic risk, political risk, and reputational risk, are often covered under these.

- Sensitivity testing

It is used to change each parameter separately and measure how much the outputs of the model fluctuate or are sensitive to different variables.

2.5 Other risks

Another risk that is quantified using LDA is credit risk, which is the estimation of expected and unexpected loss from credit defaults; see [10]. Note that [21] used the inverse Gaussian distribution to quantify credit risk and used Copula functions and Laplace transformation to run an algorithm that quantifies the corresponding probabilities. In this chapter, our focus is not on credit risk. Hence, a reader who might be interested in how the probability distribution of defaults is quantified can go through the articles by [10, 18].

3. Methodology

3.1 Introduction

In the previous section, we outlined numerous articles and textbooks that discussed various aspects of operational risk using loss distributions. Those articles and textbooks formed a literature review that helped us figure out how to quantify operational risk using loss distributions which is the main objective of our research work. Firstly though, we provide detailed description of the EVT, VaR, and their corresponding properties.

Under EVT approach, [2] explained that the analysis of the tail area of the distribution is the main focus as well as using appropriate methods for modeling extreme losses and their impact in insurance, finance, and quantitative risk management. We fit classical distributions to the data, using the maximum likelihood criteria, starting from light-tailed distributions (e.g. Weibull distribution), to medium-tailed distributions (e.g. log-normal). Under this approach, the Kolmogorov-Smirnov (KS) test and the Anderson-Darling (AD) test are adapted to measure the distance between the empirical and theoretical distribution functions only in the tail area, after deciding on the desired quantile. Readers are referred to [25, 26] for the KS test and for the AD test. Note that mean-excess plots are used to assess the validity of modeling the tails, while the Hill method is also used to get rough estimates of the shape of the parameter of a distribution; see [2]. In addition, [2] stated that the KS and AD tests can be used to examine the goodness-of-fit of models that we want to fit on the data. These tests can be used to determine which loss distribution best fits our operational loss data. These tests use different measures of discrepancy between fitted continuous distributions and empirical distributions. KS test is the best at measuring the discrepancy around the median, while the AD test is good at measuring the discrepancy for the tails.

According to [2], VaR is the largest loss an investment portfolio might sustain over a specific length of time. The time frame can be a single day, a month, a quarter, or even an entire year. According to [2], it is the $(1\text{-}\alpha)^{\text{th}}$ percentile of the loss distribution over a desired time frame, where $(1\text{-}\alpha)$ is the level of confidence and practitioners typically put it at 99.99%. Also note that [27] defined VaR as a number that indicates how much a financial institution can lose with probability over a specific time horizon, and that its measurements can be used in risk management, the assessment of risk takers' performance, and for regulatory requirements.

The method of moments is a different analytical method for quantifying risk. In this method, the mean and variance of input distributions defined at the task level are utilized to calculate the moments of the probability distribution corresponding to the task completion date. The major moments of work breakdown structure simulations may be determined using this method almost instantly and precisely, and it is still utilized in the cost risk analysis community; see [28].

Now in this methodology section, we intend to use a dataset to illustrate to readers how to quantify real-life operational risk using loss distributions. In addition, we intend to measure the descriptive statistics such as the mean, skewness, and kurtosis to obtain a general idea of how each of the considered datasets are distributed. For instance, **Tables 3** and **4** below give a summary of how skewness and kurtosis are used to give an idea of how the dataset(s) may be distributed. As indicated in [2], there are important statistical approaches to consider when running a goodness-of-fit test to a dataset, i.e. the KS and AD test. More importantly, the maximum likelihood estimation (MLE) or method of moments shall be used to estimate the model parameters.

The pivotal role of our study is to determine the best loss distribution that fits the considered dataset. To complete this role, we ought to compare between Pareto, gamma, Weibull, log-normal, Burr, and exponential distributions and investigate using specific metrics which distribution best fits our data. We are going to perform these tasks with the help of R software (the dataset used and R codes can be requested from the authors) using packages, such as moments and fitdistrplus. Having determined which distribution best fits our data, we shall use the best-fitting loss distribution to calculate the probability of loss for that specific dataset. Consequently, it may happen that the data do not seem to fit well with any distribution; then in such an instance, we would conclude that using a nonparametric approach would be of better benefit.

3.2 Goodness-of-fit test

Goodness-of-fit tests are useful to determine the validity of a theoretical model. There are different types of tests that can be used to perform the goodness-of-fit test, e.g. Kuiper, Cramer von Mises, and Pearson's chi-square test, but in this research work, we mainly focus on the KS and AD tests.

Skewness	**Data shape**
Zero	Asymmetry
Negative	Left-tailed
Positive	Right-tailed

Table 3.
Value of skewness implications.

Kurtosis	**Tail heaviness**
Zero	Equal to normal curve
Negative	Lighter-tailed
Positive	Heavy-tailed

Table 4.
Values of kurtosis implications.

3.2.1 Kolmogorov-Smirnov test

This test captures the deviation or variance around the median of the data. The KS test is computed using the maximum vertical distance between $F_n(x)$ and $F(x)$, where $F_n(x)$ is the CDF of the empirical formula and $F(x)$ is the CDF of the observed data.

According to [2], KS test statistic is computed as follows: let D^+ be the largest difference between $F_n(x)$ and $F(x)$ and let D^- be the largest value between $F(x)$ and $F_n(x)$. Then, mathematically,

$$D^+ = sup_x\{F_n(x) - F(x)\} \tag{4}$$

$$D^- = sup_x\{F(x) - F_n(x)\}. \tag{5}$$

Thus, the KS statistic is calculated as:

$$KS = \sqrt{n}\ max\ \{D^+, D^-\}, \tag{6}$$

which can be written as,

$$\sqrt{n}\,max\left\{sup\left\{\frac{j}{n} - z_{(j)}\right\}, sup\left\{z_{(j)} - \frac{j-1}{n}\right\}\right\}. \tag{7}$$

where n is the number of observations, $z_{(j)} = F\left(x_{(j)}\right)$ and $j = 1,2, \dots ,n$.

For hypothesis testing, the null hypothesis is that the dataset that we will use for illustration purpose "Taxi claims" data follows the specified distribution, and the alternative hypothesis is that the "Taxi claims" data does not follow the specified distribution using a critical value of 5% throughout.

3.2.2 Anderson-Darling test

This test is best suited for computing discrepancies around the tails. The test is mostly used for heavy-tailed data, and the test statistic of the AD test is given by:

$$\text{AD} = \sqrt{n}\,sup\left|\frac{F_n(x) - F(x)}{\sqrt{F(x)(1 - F(x))}}\right|. \tag{8}$$

The computing formula is given by:

$$\text{AD}^2 = -n + \frac{1}{n}\sum_{j=1}^{n}(1 - 2j)\log\left(z_j\right) - \frac{1}{n}\sum_{j=1}^{n}(1 - 2(n - j)\log\left(1 - z_j\right). \tag{9}$$

3.3 Sensitivity analysis

In this section, we perform the sensitivity analysis of the distribution that will be fitted to our data. The purpose of doing this is to show the effect of the different parameters' behavior on the tail and peak portion of the probability distribution. When testing for the effect of a parameter, we will fix all other variables and vary the parameter of interest.

In **Figure 1**, we varied λ as 0.5, 1 and 1.5. We can clearly see that at 0.5 we had a thicker tail, while at 1.5 we observe a thin tail. Thus, the more we increase λ we obtain

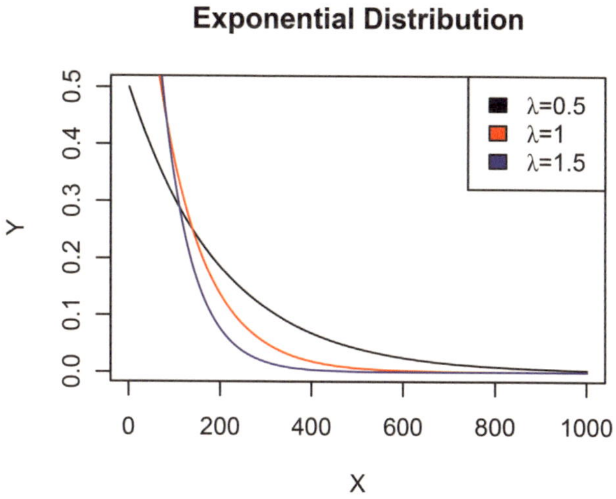

Figure 1.
Exponential distribution.

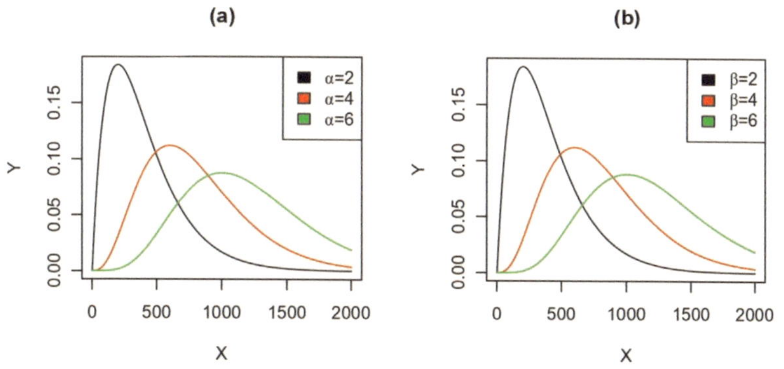

(a) β=2 with $\alpha \in \{2,4,6\}$	(b) α=2 with $\beta \in \{2,4,6\}$

Figure 2.
Gamma distribution.

an exponential distribution with a thinner tail. In **Figure 2a**, β is fixed to be 2 with $\alpha \in \{2, 4, 6\}$, and we can clearly see that when α =2, the resulting gamma distribution has thin tails; however, for large α, the distribution has thicker tail. In **Figure 2b**, α is fixed and β is varied. It is observed that the gamma distribution has thin tails, when β is small and we have thick tails for large β.

In **Figure 3a**, with the β fixed, for small α, the Weibull distribution has a thick tail and a thin tail for large values of α . In **Figure 3b**, given that α is fixed, as β increases, we observe thicker tails; however, for small β, we observe thin tails. For β fixed, in **Figure 4a**, a small α yields a thicker tail; however, a larger one yields thin tails. Next in **Figure 4b**, with α fixed; a small β yields thin tail while a large one yields thicker tails.

In **Figure 5a**, when α is fixed at 0.5 and γ to be 10, we varied β to be 10, 20, and 30, and it is observed that when β is 30 there is a thicker tail and when β is 10 there is a

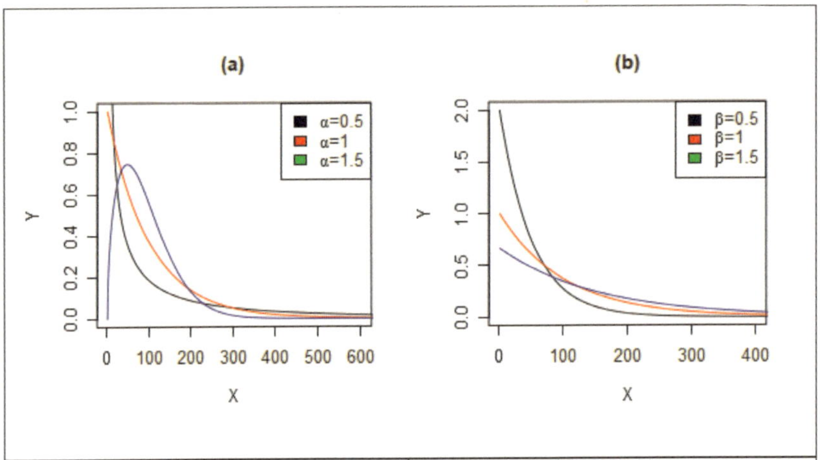

(a) β=1 with $\alpha \in \{0.5, 1, 1.5\}$ | (b) α=1 with $\beta \in \{0.5, 1, 1.5\}$

Figure 3.
Weibull distributions.

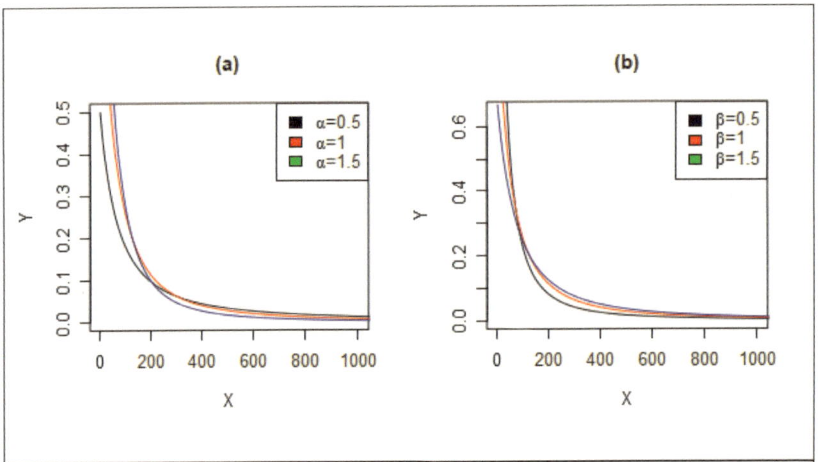

(a) β=1 with $\alpha \in \{0.5, 1, 1.5\}$ | (b) α=1 with $\beta \in \{0.5, 1, 1.5\}$

Figure 4.
Pareto distributions.

thin tail. That is, when the value of β is increased, the tails become thicker, and the opposite is true. In **Figure 5b**, α is fixed to be 0.5 and β to be 50, γ is varied to be 5, 10, and 15, and it is observed that when γ is 5 there is a thicker tail and when γ is 15 there is a thin tail. When we increase the value of γ the tails become thinner, and the opposite is true. In **Figure 5c**, γ is fixed at 10 and β to be 25, α is varied to be 0.5,1, and 1.5 and when α is 1 there is a thicker tail and when α is 1.5 there is a thin tail. When we increase the value of α, the tails become thinner, and the opposite is true.

In **Figure 6a**, we fixed the mean at 1 and varied standard deviation (stdev) to be 0.5, 1, and 1.5, and it is observed that when the stdev is 1, there is a thicker tail and when stevd is 0.5 there is a thin tail. In **Figure 6b**, we fixed stdev to be 1 and varied

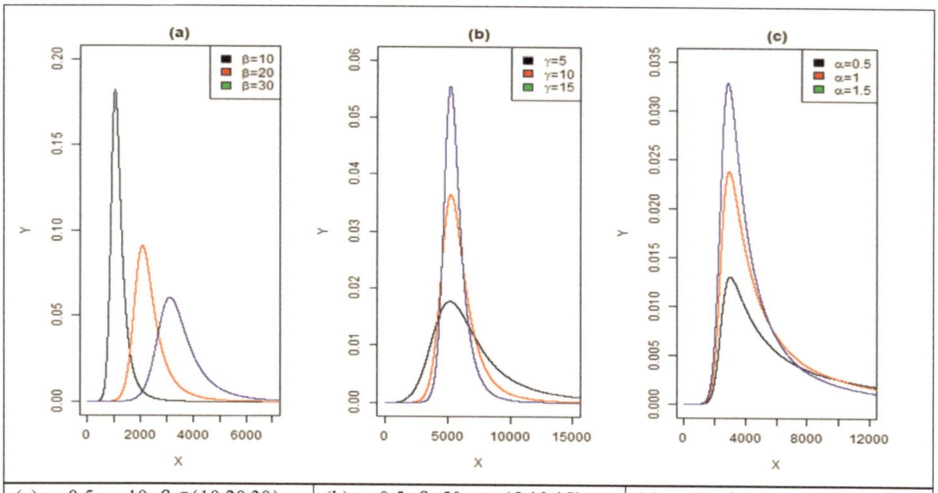

(a) α=0.5, γ=10, $\beta \in \{10,20,30\}$ | (b) α=0.5, β=50, $\gamma \in \{5,10,15\}$ | (c) γ=10, β=25, $\alpha \in \{0.5,1,1.5\}$

Figure 5.
Burr distributions.

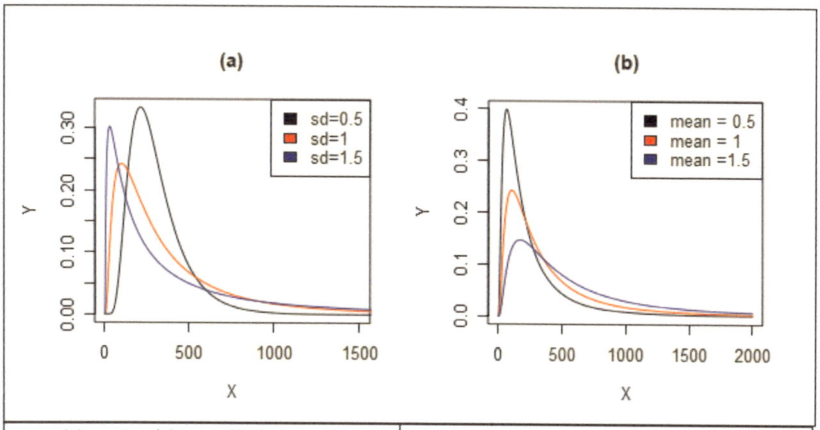

(a) μ=1 with $\sigma \in \{0.5, 1, 1.5\}$ | (b) σ=1 with $\mu \in \{0.5, 1, 1.5\}$

Figure 6.
Log-normal distributions.

the mean to be 0.5, 1, and 1.5. We can clearly see that when the mean is 1.5, there is a thicker tail and when mean is 0.5 there is a thin tail.

3.4 Analysis of Taxi claims data

Again, the Taxi claims data and R codes used to analyze the data can be obtained from the authors on request. **Table 5** provides the descriptive summary of the Taxi claims data.

Both the calculated skewness and kurtosis are positive, thus based on the information based on **Tables 3** and **4**, we can conclude that the dataset is skewed to the right and is heavy-tailed.

Mean	**13232.41**
Standard deviation	28415.63
Median	4500
Skewness	6.474064
Kurtosis	63.63799

Table 5.
Descriptive statistics of the Taxi claims data.

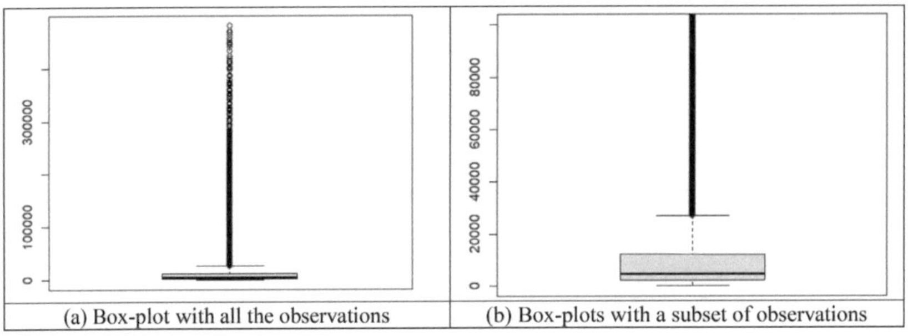

Figure 7.
Boxplot of Taxi claims data.

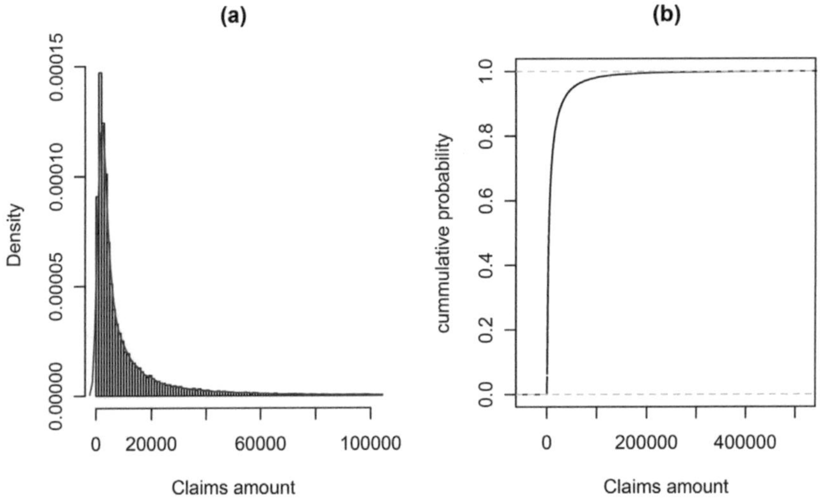

Figure 8.
Histogram and CDF of taxi claims data.

From **Figure 7a**, it is observed that the Taxi claims data has many extreme observations on the right tail because **Figure 7b** that excludes extreme values shows that the lower quantile of the boxplot. Next, **Figure 8** provides the corresponding PDF via a histogram (see **Figure 8a**) and the CDF (see **Figure 8b**) of the Taxi claims data.

Next, we fit a gamma, Weibull, log-normal, Burr, and Pareto distributions to the Taxi claims data. The parameter estimates and goodness-of-fit test results are provided in **Figure 9**.

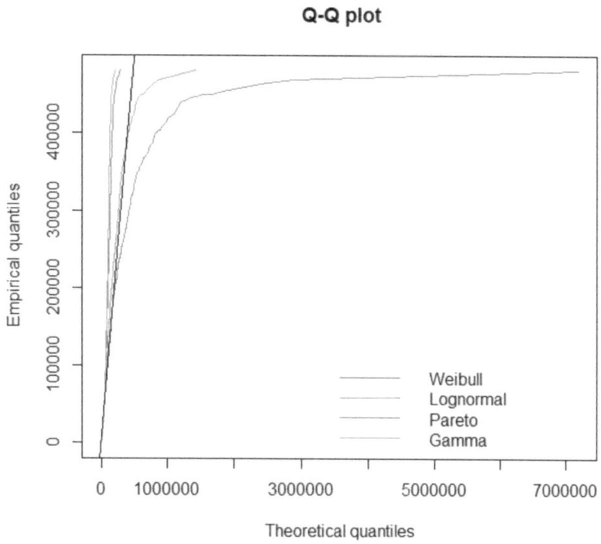

Figure 9.
QQ plots of the Taxi claims data fitted for different loss distribution.

Distributions	Pareto	Gamma	Weibull	Log-normal
Parameters	$\alpha = 1.7596$ $\beta = 10600.4071$	$\alpha = 0.64547767$ $\beta = 0.00004879$	$\alpha = 0.7230$ $\beta = 10097.898$	$\mu = 8.5437$ $\sigma = 1.3211$
AD test value	338.3246	1727.9785	1040.7859	115.3129
KS test value	0.06150089	0.1356	0.0886	0.0410

Table 6.
MLE parameters and goodness-of-fit values.

From **Figure 9**, it is observed that the log-normal is a better fit than the other corresponding distributions (**Table 6**).

It is observed that the log-normal distribution has the lowest KS and AD values out of all the tested distributions, and we can assume that the median of our data and tail is best explained by the log-normal distribution. Overall, we can see that the heavy-tailed distributions, log-normal, and Pareto distribution fit our data very well, and the thin-tailed distributions, gamma and Weibull distributions fit is poor. We can conclude that our data is best modeled by the log-normal distribution.

Figure 10 is the Q-Q plot for the log-normal distribution, although the log-normal distribution was the best distribution to model the Taxi claims data according to the AD test and the KS test, it was not good for modeling extreme losses or extreme values (see the tails of **Figure 10**).

3.5 VaR sensitivity

The maximum amount that can be lost during a specific holding period with a given level of confidence is known as VaR (i.e. value-at-risk). Four methods will be used to determine VaR:

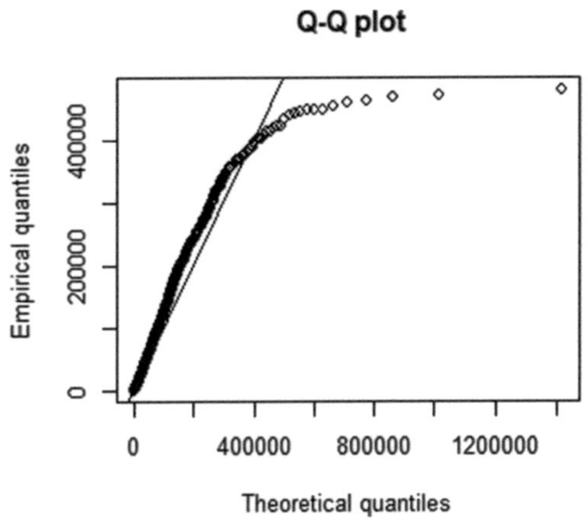

Figure 10.
Log-normal QQ-plots.

The empirical approach—it entails sorting the data from lowest to highest and taking quantiles. Here, the focus is on the 90th, 95th, 97.5th, and 99th percentiles.

The parametric approach—it entails the use of the fitted model. Here the log-normal was the best distribution for the Taxi claims data. Thus, to calculate the VaR, we used a "VaRes" package on R and put in the shape and scale parameter and computed the 90th, 95th, 97.5th, and 99th percentiles.

The stochastic approach—it entails simulating from a log-normal distribution with shape and scale parameter found from Taxi claims data and computing the 90th, 95th, 97.5th, and 99th percentiles.

The generalized extreme value (GEV) distribution approach—it entails simulating from the GEV distribution and computing the 90th, 95th, 97.5th, and 99th percentiles.

Table 7 and **Figure 11** gives a summary of the different approaches of calculating VaR. Note that the column "$VaR_{0.9}$" in **Table 7** under empirical approach can be interpreted as follows: "we can be 90% confident that the maximum amount claimed will be R30 616.67." In the case of $VaR_{0.9}$, it is evident that the empirical VaR are close to the extreme value approach, and as the percentiles increase from $VaR_{0.975}$ to $VaR_{0.99}$, the GEV distribution was overestimating the risk. Thus, for the Taxi claims data, one would be more inclined to use the empirical approach because it does not assume any underlying distribution as the log-normal distribution seem to be poor in capturing the extreme tail component of the data. The rest of the amounts can be interpreted in the same way for the corresponding percentage levels.

Approach	$VaR_{0.9}$	$VaR_{0.95}$	$VaR_{0.975}$	$VaR_{0.99}$
Empirical approach	30616.67	52515.09	83327.03	139690.1
Parametric approach (log-normal)	27911.04	45105.2	68395.31	110980.1
Stochastic	27981.36	45163.61	68196.93	109861.3
Extreme value	30009.28	59051.74	114121.5	269268.2

Table 7.
VaR for the Taxi claims data.

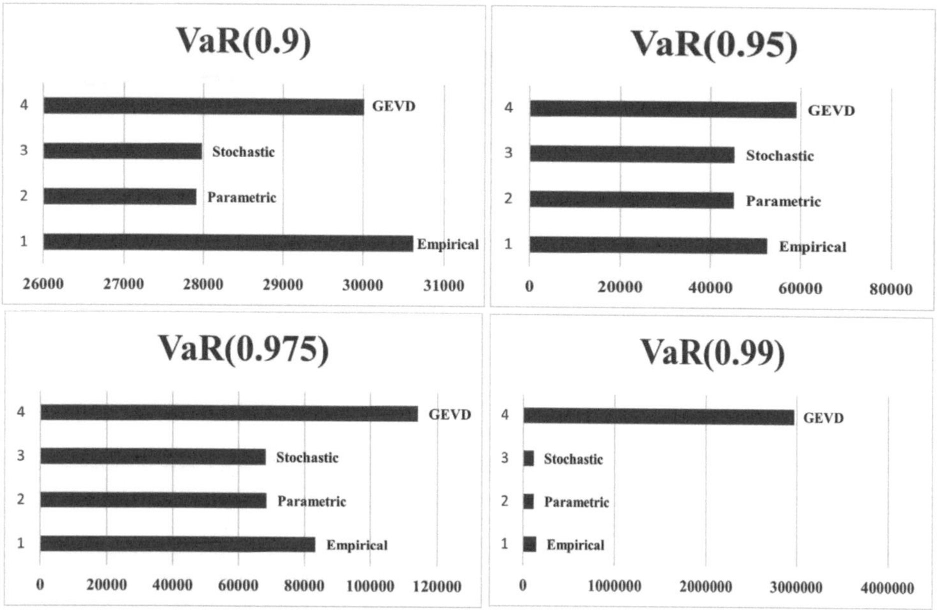

Figure 11.
Graphical representation of VaR.

In this section, we aimed to learn how to quantify operational risk using parametric loss distributions (i.e. exponential, log-normal, gamma, Weibull, Pareto, and Burr distributions). As an example, for the chapter application, we fitted all the distributions to the Taxi claims data. Overall, we can conclude that out of all the distributions fitted, the log-normal distribution seemed to be the best-fitting distribution. We also observed some of the drawbacks of using the parametric distributions in that it tends to fail in capturing the tail as well as the peak of the data very well and makes one think maybe the true underlying distribution could be different from the fitted one. This means that if a parametric distribution method is applied to quantify operational risk, it is better to fit different distributions to the tail and the body to get better estimates. This is evident in our data where after a certain threshold, it is observed that the log-normal distribution underestimates the probability in the tail area; however, the GEV distribution fits better.

4. Conclusion

We have discovered that there are many methods used in quantifying operational risk. Therefore, it is proven handy for a risk analyst or anyone that is working in the risk analysis department of any institution to possess vast knowledge of multiple statistical concepts, and methods to apply in any given situation as each risk requires a different quantification approach. The Taxi claims data we have analyzed provides a good foundation for analyzing operational risk, but it does not represent all possible situations one might find in real life. Because Taxi claims are limited to the highest replacement value of a Taxi; in our case, we have realized that our data does not contain very extreme values; therefore, high losses are limited. The possibility of ruin would likely be due to the risk class of "high frequency and low magnitude." Using the

Taxi claim data, we have provided a good example of how operational risk may be quantified and observed that the log-normal distribution is a better fit in this case. Nevertheless, it failed to model the extreme values of the Taxi claims data. Hence, the GEV distribution can be used to model those extreme values. However, in our research work, we did not dwell much on the GEV distribution. We used four approaches to calculate VaR in our analysis.

According to existing empirical evidence, the overall pattern of operational loss severity data is characterized by significant kurtosis, severe right-skewness, and a very heavy right tail caused by multiple outlying incidents. Fitting some of the common parametric loss distributions such as Weibull, log-normal, Pareto, gamma distributions, and so on is one way to calibrate operational losses. One disadvantage of utilizing these distributions is that they may not suit both the centre and the tails perfectly. Mixture distributions may be explored in this scenario. EVT can be used to fit a GPD to extreme losses surpassing a high predetermined threshold. The characteristics of the GPD distributions which are derived using the EVT approach are very sensitive to extreme data and the choice of threshold, which is a drawback of this approach.

The disadvantages of using loss distributions are that they do not model well many datasets in the presence of outliers and extreme values. Consequently, [2]'s Chapters 7 and 8 discuss other possible replacements for the loss distributions, e.g. alpha-stable distributions, GEV distribution, and GPD. The latter two distributions are part of EVT family of distributions.

Overall, it appears from the literature study done for this chapter that operational risk managers are focused on developing a model that would accurately represent the likelihood of the tail occurrences and producing a model that would realistically account for the probability of losses reaching a large amount is essential. Because, the latter is important for estimating the VaR.

Since we mainly focused on parametric distributions, and it might be interesting to use the nonparametric approach to quantify risk and some other mixture of distributions method. Therefore, this means that some of the possible additions to research work can be looking at more complex statistical distributions with better tail capturing ability. We also looked at multiple ways of finding VaR and from our findings, and we noticed that the empirical approach was the better way of quantifying the Taxi claims data's risk—note though the conclusion reached is data-dependent, which means it that a different conclusion will be made for a different dataset.

Acknowledgements

We would like to thank the University of the Free State for their continued support and assistance.

Author details

Retsebile Maphalla, Moroke Mokhoabane, Mulalo Ndou and Sandile Shongwe*
Faculty of Natural and Agricultural Sciences, Department of Mathematical Statistics and Actuarial Science, University of the Free State, Bloemfontein, South Africa

*Address all correspondence to: shongwesc@ufs.ac.za

References

[1] Vaughan E, Vaughan T. Fundamentals of Risk and Insurance. New Jersey: John Wiley & Sons; 2003

[2] Chernobai A, Rachev S, Fabozzi F. Operational Risk: A Guide to Basel II Capital Requirements, Models, and Analysis. New Jersey: John Wiley and Sons, Inc; 2007

[3] Sweeting P. Financial Enterprise Risk Management. New York: Cambridge University Press; 2011

[4] Wilson T. Portfolio credit risk. Economic Policy Review. 1998;**4**(1): 71-82

[5] BIS. Basel II: International Convergence of Capital Measurement and Capital Standards: Framework. BIS; 2006

[6] Byatt G. Why and how we should quantify risk. Principal Consultant. Risk Insight Consulting [Internet]. 2019. Available from: https://irp-cdn.multiscreensite.com/8bbcaf75/files/uploaded/Risk%20Insight%20-%20why%20and%20how%20we%20should%20quantify%20risk%20-%20full.pdf. [Accessed: July 25, 2022]

[7] Bank of International Settlements. Working Paper on the Regulatory Treatment of Operational Risk [Internet]. 2001. Available from: www.BIS.org [Accessed: May 09, 2022]

[8] Allen L, Boudoukh J, Saunders A. Understanding Market, Credit, and Operational Risk: The Value-at-Risk Approach. Oxford: Blackwell Publishing; 2004

[9] Shih J, Samad-Khan A, Medapa P. Is the Size of an Operational Loss Related to Firm Size? [Internet]. 2000. Available from: http://www.stamfordrisk.com/docs/Is_the_Size_of_an_Operational_Loss_Related_to_Firm_Size_(Jan_00).pdf. [Accessed: May 5, 2022]

[10] Maccario A, Sironi A, Zazzaea C. Credit risk models: An application to insurance pricing. SSRN Electronic Journal. 2003;**384380**

[11] Calderín-Ojeda E. A Note on parameter estimation in the composite Weibull–Pareto distribution. Risks. 2018; **6**(1):11. DOI: 10.3390/risks6010011

[12] Rebello A, Martinez A, Goncalves A. Analytical solution of modified point kinetics equations for linear reactivity variation in subcritical nuclear reactors adopting an incomplete gamma function approximation. Natural Science Journal. 2012;**4**(1):919-923

[13] Ahmad Z, Mahmoudi E, Hamedani GG. A family of loss distributions with an application to the vehicle insurance loss data. Pakistan Journal of Operational Research. 2019; **15**(3):731-744

[14] Ahmad A. Bayesian Analysis and Reliability Estimation of Generalized Probability Distributions. Balrampur: AIJR Publisher; 2019

[15] Bolviken E, Haff I. One Family, Six Distributions-A Flexible Model for Insurance Claim Severity. Oslo, Norway: Department of Mathematics, University of Oslo; 2018. DOI: 10.48550/arXiv.1805.10854

[16] Burnecki K, Misiorek A, Weron R. Loss Distributions. Munich Personal RePEc Achive. Wrocław, Poland: Hugo Steinhaus Center, Wrocław University of Technology; 2010

[17] Cizek P, Hardle W, Weron R. Statistical Tools for Finance and Insurance. Berlin: Springer; 2005

[18] Frachot A, Georges P, Roncalli T. Loss distribution approach for operational risk. SSRN Electronic Journal. 2001;**1032523**:2-12

[19] Hakim A, Fithriani I, Novita M. Properties of Burr distribution and its application to heavy-tailed survival time data. Journal of Physics. 2021;**1725**(1): 012016

[20] Jamal F, Chesneau C, Nasir M, Saboor A, Altun E, Khan M. On A modified Burr XII distribution having flexible Hazard rate shapes. Mathematica Slovaca. 2019;**70**:1. DOI: 10.1515/ms-2017-0344

[21] Schonbucher P. Taken to the limit: Simple and No so simple loan loss distribution. SSRN Electronic Journal. 2002:23. DOI: 10.2139/ssrn.378640. Available from: https://ssrn.com/abstract=378640

[22] Supriatna A, Parmikanti K, Novita L, Sukono., Betty, S. and Talib, A. Calculation of Net Annual Health Insurance Premium Using Burr Distribution. Paris: IEOM Society International; 2018

[23] Murphy K, Carter C, Brown S. The exponential distribution: The good, the bad and the ugly. A practical guide to its implementation. Annual Reliability and Maintainability Symposium. 2002 Proceedings. Seattle, WA, USA: IEEE. 2002. DOI:10.1109/RAMS.2002.981701

[24] Valencia A, Jaramillo W. Quantifying operational risk using the loss distribution. In: Proceedings of the Seventh European Academic Research Conference on Global Business. Zurich, Switzerland: Economics, Finance and Banking (EAR17 Swiss Conference); 2017. ISBN: 978-1-943579-46-4. Paper ID: Z702

[25] Eadie W, Drijard D, James F, Roos M, Sadoulet B. Statistical Methods in Experimental Physics. Amsterdam: Elsevier Science Ltd; 1983

[26] Anderson T, Darling D. Asymptotic theory of certain goodness of fit criteria based on stochastic processes. Annuals of Mathematical Statistics. 1952;**23**(1): 193-212

[27] Manganelli S, Engle F. Value at risk model in finance. European Central Bank Working Paper Series. 2001;**75**(1):1-40

[28] Covert R. Using Method of Moments in Schedule Risk Analysis. MCR, LLC; 2011. DOI: 10.1002/9780470400531.eorms0591